Springer Proceedings in Geography and Sustainability

Springer Proceedings in Geography and Sustainability features state-of-the-art contributions from conferences, workshops, and similar meetings. The series presents a snapshot of timely developments in fields including but not limited to physical, human, and integrated geography; cross-disciplinary approaches to environmental challenges; and sustainability writ large. It aims to foreground trends emerging from meetings around the world, especially regions that are traditionally underrepresented.

Proposals must include the following:

- A description of the meeting's aims and theme
- The name, place, date, and website of the meeting
- The committee(s) and/or organization(s) responsible for planning the meeting
- A description of the organizer's review process for submissions

Please reach out to the publishing editor, Andrea Brody-Barre (andrea.brody-barre@springer.com), with enquiries.

Nixon Ronoh • Charles Rono
Editors

Rethinking Our Ecosystems

Proceedings of the International Conference on Renewable Energy, Climate Change and Ecosystem Restoration 2024

Springer

Editors
Nixon Ronoh
School of Sciences and Aerospace Studies
Moi University
Eldoret, Kenya

Charles Rono
Department of Chemistry
Kisii University
Kisii, Kenya

ISSN 3059-3344 ISSN 3059-3352 (electronic)
Springer Proceedings in Geography and Sustainability
ISBN 978-3-031-95780-2 ISBN 978-3-031-95781-9 (eBook)
https://doi.org/10.1007/978-3-031-95781-9

This Springer imprint is published by the registered company Springer Nature Switzerland AG
The registered company address is: Gewerbestrasse 11, 6330 Cham, Switzerland

This publication of conference proceedings is dedicated to the authors, the multidisciplinary editorial teams from different institutions across the globe, restoration practitioners in the Great Rift Valley Region and the various technical teams from the County Governments in the North Rift Economic Bloc (NOREB) Region, Kenya, and Moi University. It is the individual and collective efforts of these teams that enabled the success of this project.

Foreword

Climate change has attracted considerable interest, both globally and locally, from various stakeholders' intent on solving the myriad of emerging challenges that bedevil and risk the human future. The recent past has seen many perceived solutions based on implementable mitigation and adaptation efforts. While theory and practice seek different paths of such efforts, a merger of both presents the best alternative to addressing the effects of climate change. Renewable energy solutions, ecosystem restoration and sustainability efforts work better if converged into a concerted strategic stakeholder partnership that also involves local communities, their governments and learning institutions. An international conference on climate change and renewable energy goes a long way in achieving this and offers a forum to disseminate climate action plans, through research, capacity building and knowledge enhancement amongst practitioners, policy makers, industry players, research institutions, indigenous people and other industry players. The International Conference on Renewable Energy, Climate Change and Ecosystem Restoration (ICRECC) is an annual event co-hosted by the County Government of Uasin Gishu and Moi University, whose main objective is to encourage collective effort and action towards climate change mitigation through ecosystem restoration and sustainable use of renewable energy. The conference is founded on the need to disseminate findings from constructive dialogue and engagement between stakeholders at different industry levels. While the emphasis is deeply on the landscapes that cover the Great Rift Valley, ICRECC envisages a Renewable Energy Climate Change and Ecosystem Restoration model that can be replicated in regions beyond, structured to drive global progress towards a sustainable and resilient future. The conference is designed to serve, firstly as a platform for knowledge exchange and secondly, as a forum through which result-oriented collaborative activities amongst stakeholders can be documented and shared with communities. The overall theme for the inaugural conference ICRECC-2024 'Rethinking Our Ecosystems' focuses on cross-disciplinary themes on climate change, ecosystem conservation, adaptation and sustainability efforts towards a circular economy in landscape restoration.

The authors of this volume present findings selected from ICRECC2024 conference proceedings, merging together ideas cutting across advocacy, knowledge exchange and stakeholder engagement. The chapters of the volume and the subthemes of the conference are co-aligned with sustainability efforts in mitigation and adaptation of climate change. The first subtheme was *Restoration of Landscapes*, which included restoration financing, engagement of local actors, diversity and inclusion, advocacy and awareness. The second subtheme was *Emerging Developments in STEAM Towards Sustainability in Mitigation and Adaptation to Climate Change*. The third subtheme was *Ecosystem, Biodiversity and Natural Resource Management*. The fourth and final subtheme was *Cross-Cutting Themes*, which included socio-economic aspects, sustainable agriculture, food security, health and climate smart practices.

The authors Dr. Nixon Ronoh and Dr. Charles Rono are members of the faculty of Moi University (Kenya) and Kisii University (Kenya), respectively. The editorial team is composed of Dr. Rose Ramkat, Prof. Frederick Nyamwala, Dr. Njira Njira Pili, Dr. Sarah Cherono and Dr. Charles Mutai, who are all members of the faculty at Moi University (Kenya), Dr. Selly Kimosop, Chairperson of the Public Service Board in the County Government of Elgeyo Marakwet (Kenya) and Dr. Edna Milgo, a member faculty at the Open University of Kenya.

Faculty, School of Engineering
Moi University,
Eldoret, Kenya

Charles C. Nzila

Preface

The merger between theory and practice is oftentimes never straightforward. The foundational principles of the International Conference on Renewable Energy, Climate Change and Restoration of Ecosystems (ICRECC) stemmed from the initiative to bring together researchers and practitioners to a point of convergence. ICRECC provides such a platform, where adaptation and mitigation efforts towards climate change are solidified through knowledge exchange and dissemination. The conference brings together diverse sets of stakeholders, including those directly engaged in climate action through advocacy, financing, research and innovation, with different points of application in communities around the Great Rift Region and beyond. Some of the problems that ICRECC purposes to solve include pollution, waste management and the encroachment of riparian lands. The conference firstly aims to drive research and new thought towards enhancement of vegetative cover and secondly advocates for green ventures that promote utility of renewable energy sources. While the County Government brings in community engagement, Moi University brings into ICRECC its internationally renowned research prowess. Besides helping address real-time community energy challenges, this volume will play a greater part in documenting and disseminating feasible solutions towards adaptation and mitigation strategies for climate change. This inaugural ICRECC publication is sectioned into three chapters, namely "The Impact of Climate Change", "Restoration of Ecosystems" and "Adaptation Strategies". Subsequent volumes will include topics on renewable energy and cross-cutting themes.

Eldoret, Kenya — Nixon Ronoh
Kisii, Kenya — Charles Rono

Introduction

Integrated ecosystem-based approaches to climate change mitigation and adaptation towards climate-resilient communities and ecosystems beyond the Great Rift Valley.

Over the past few decades, mankind and the ecosystems in general have experienced severe threats from the accumulated effects of climate change as scientifically projected over a century ago. Early observations showcased approximations that human activities such as burning fossil fuels and degradation of forests significantly increased the atmospheric accumulation of greenhouse gases, mainly carbon dioxide and other infrared absorbing gases. For instance, Arrhenius' and other scientists' pioneering projections over a century ago on the increased global temperatures in 30 centuries due to the effect of doubled accumulation of greenhouse gases has already been realized in the last few decades, which is barely a century ago and the impact is continuously on the rise.[1,2] Since then, a lot of climate scientists have made contributions to the understanding of human-induced changes in climate throughout the globe. Accelerated industrialization over the past five decades in various parts of the continents brought about significant rise in the levels of carbon dioxide and other greenhouse gases in the atmosphere. This has consequently led to the rise in the average global temperatures over the years, and the impact of these global temperatures is currently on a scale that is threatening our ecosystems.

The title of the book *Rethinking Our Ecosystems* draws inspiration from the main theme of the conference 'Restoration of Our Landscapes' which targeted the various restoration programmes of the landscapes in the Great Rift Valley and beyond. Considering the major challenges facing humanity today from climate change and the downstream implications of degraded ecosystems, a contemporary science and holistic approach towards solving the climate change crisis is necessary. Such contemporary solutions not only consider multidisciplinary efforts but also promote

[1] Crawford E. Arrhenius' 1896 model of the greenhouse effect in context. AMBIO-STOCKHOLM-. 1997 Feb 1;26:6–11.http://www.jstor.org/stable/4314543

[2] Anderson TR, Hawkins E, Jones PD. CO2, the greenhouse effect and global warming: from the pioneering work of Arrhenius and Callendar to today's Earth System Models. Endeavour. 2016 Sep 1;40(3):178–87.https://doi.org/10.1016/j.endeavour.2016.07.002

cross-sectoral strategies for concerted efforts towards climate-resilient communities and ecosystems.

Climate change is an umbrella definition for notable variations in the long-term average weather conditions of a place. The 2010 National Climate Change Response Strategy (NCCRS) considered climate change as one of the most serious threats to sustainable development locally and globally. About 90% of all natural disasters that have severely hit humankind are due to extreme climate change events, with secondary hazardous effects being felt in key sectors such as health, food security, natural resources and infrastructural developments.[3] According to NCCRS, the Kenyan population became susceptible to malaria and other vector-borne diseases, a phenomenon largely attributed to adverse temperature conditions.

Diverse climate change actors and practitioners have devised an array of strategies and measures to mitigate climate change through ecosystem restoration programmes. With the progressive growth of urbanization programmes globally, the built environment for a long time has focused a lot on architectural landscapes at the expense of natural landscapes.[4] The recent scholarly contributions in the built environment focus and advocate for a paradigm shift to an integrated urban and natural environments, considerations of architectural landscapes together with the natural landscapes in the built process and bridging the gaps of divisions in the studies of the cities from the studies of land. According to Alexandra and Riddington, it is well demonstrated that the evolution of natural landscapes depends largely on the dynamic shifts in human social-cultural and economic behaviour. These proceedings' took advantage of the now established importance of human social behaviour and cultures in the integrated restoration and conservation of natural landscapes and hence the benefit in ecosystem investments.[5]

In these proceedings, the thematic work complements and fills the existing divisions in rethinking and restoration of our ecosystems from four fronts: Firstly, climate change mitigation through restoration of landscapes. Secondly, it considers emerging developments in science, technology, engineering, arts and mathematics (STEAM) towards sustainability in mitigation and adaptation to climate change. STEAM particularly features emerging developments on creativity and imaginative aspects based on science, technology, engineering and mathematics. The work goes beyond the existing ideologies, scholarship and practices in that it considers data-driven approaches and initiatives integrating artificial intelligence techniques and the Internet of things as integral in achieving sustainability of adaptation measures, innovations and mitigation strategies to climate change. Thirdly, this work explores

[3] National Climate Change Response Strategy Government of Kenya (NCCRS) https://cdkn.org/sites/default/files/files/National-Climate-Change-Response-Strategy_April-2010.pdf. Accessed on 15th August 2024.

[4] Way T, Oaks D, Karmon D, Buckley JM, Cunningham TA, Ding L, Gandy M, Hossain L, Jacobs S, Liu M, Melnick RZ. Rethinking the urban landscape. https://www.repository.cam.ac.uk/handle/1810/362001

[5] Strassburg BB, Iribarrem A, Beyer HL, Cordeiro CL, Crouzeilles R, Jakovac CC, Braga Junqueira A, Lacerda E, Latawiec AE, Balmford A, Brooks TM. Global priority areas for ecosystem restoration. Nature. 2020 Oct 29;586(7831):724–9. https://doi.org/10.1038/s41586-020-2784-9

nature-based solutions in the restoration of ecosystem, biodiversity and natural resource management. Fourth and finally, the proceedings considered ideologies from diverse and cross-cutting themes that incorporate sustainable agriculture, food security, health and other climate smart practices. It is considered that the advancement of technology, socio-economic indices and evolution of human cultures are directly impacting the evolution of natural landscapes. By exploring opportunities in this space, practitioners and other cross-sectoral actors will likely develop improved ethos of ecosystem restoration and new policies where collective action and conservation of natural landscapes dominate.

This volume showcases a collection of selected peer-reviewed work from practitioners to research advances in academia and industry. It sets out to describe focused ecosystem restoration programmes that promote mitigation and adaptation strategies towards climate-resilient communities and ecosystems.

It has been structured into three broad areas: Part I showcases some of the evidence-based impact of climate change within and beyond the Great Rift Valley, Part II is a collection of papers aimed at restoration of ecosystems and mitigation strategies and Part III is a collection of papers featuring adaptation strategies.

Faculty, School of Sciences and Aerospace Studies Nixon Ronoh
Moi University, Eldoret, Kenya
Faculty, School of Pure and Applied Sciences Charles Rono
Kisii University, Kisii, Kenya

Acknowledgement

The volume corresponding editors acknowledge the valuable input of the editorial desk, namely

1. Dr. Rose Ramkat
2. Prof. Dr. Frederick Nyamwala
3. Dr. Njira Njira Pili
4. Dr. Sarah Cherono
5. Dr. Edna Milgo
6. Dr. Charles Mutai
7. Dr. Selly Kimosop

The editors further appreciate the efforts of the conference logistic teams led by Naomi Nkonge of Moi University, Dorothy Jerono of Save Kenya Water Towers and Kichumba Barno and Phillip Lagat of the County Government of Uasin Gishu.

Contents

About the Authors

Nixon Ronoh, PhD, is a faculty member, School of Sciences and Aerospace Studies, Moi University, Kenya. He has a specific interest in statistical estimation tools for solving multidimensional sampling problems. Currently, Nixon works with a research team studying the applications of artificial intelligence technologies in innovative and sustainable strategies towards renewable energy solutions and the adaptation and mitigation of climate change.

Charles Rono, PhD, is a researcher and a faculty member in the Department of Chemistry, Kisii University. He is actively involved in collaborative interdisciplinary research focused on sustainable science solutions to climate change and ecosystem health. Some of his contributions include harnessing data chemistry and artificial intelligence for sustainability in product and process development and phytochemicals in the development of materials for renewable energy applications.

Impact of Climate Change

Climate Change-Induced Cost on Soil Security and Its Implication on Food Security in Homa-Bay County, Kenya

Wicklife Ojallah and Janet Korir

Abstract

While there are studies on the effect of climate change on food insecurity and coping strategies among fishing communities in Homa Bay County, studies relating to the effect of climate change on soil quality and food security aren't evident. This paper detailed influence of climate change on soil security and its implication on food security in the study area. It examined the cost of anthropogenic-based practices (ABPs) on climate change and interrogated their ripple effect on soil and food security situation. Praxeology theory was used to explore environmental consequences of ABPs. Two hundred and fifty respondents were sampled from 262,036 households for quantitative data and 20 key informants purposively identified for qualitative data. It was a cross-sectional survey study and quantitative data were analyzed descriptively and inferentially using Statistical Package for Social Sciences (SPSS) version 25.0. Opinions from FGDs, KII, and field observation were qualitatively analyzed. Results showed monoculture (96.2%), tractor tillage (71.4%), livestocking (82.3%), stone mining (92.4%), deforestation (89.1%), and hill slope-based practices (87.9%) induced climate change. Binary logistic regression analysis (BLRA) indicated deforestation (B = −0.462, S.E = 0.381, and P < 0.032), stone mining (B = −0.756, S.E = 0.364, and P = 0.038), conventional tillage (B = −0.961, S.E = 0.489, and P = 0.05), and slope-based activities (B = −0.801, S.E = 0.368, and P = 0.0290) predicted food insecurity. Study concludes that ABPs result to climate change, soil insecurity and significantly cause food insecurity in Homa-Bay County. The

W. Ojallah (✉) · J. Korir
Department of Geography and Environmental Studies, School of Arts and Social Sciences, Moi University, Eldoret, Kenya

N. Ronoh, C. Rono (eds.), *Rethinking Our Ecosystems*, Springer Proceedings in Geography and Sustainability, https://doi.org/10.1007/978-3-031-95781-9_1

study recommends suspension of slope-based activities on elevated areas particularly on Homa hills and review stone extraction, and tractor cultivation policies, and intensive and purposive reforestation programs.

Keywords
Climate change · Induced cost · Soil security · Food security

1 Introduction

While the cost of climate change is a reality touching almost all aspects of human life [25] and being a subject that has attracted a wide range of scholarly attention coupled with heavy funding both at the micro and macro scale [34], the subject still remains unresolved [42]. However, all is not lost for there is a glittering hope evident through the collective climate-change actions that can lead to constructive hope [40]. It is against this solid hope that this paper intended to examine the cost of climate change on soil security and its implication on food security particularly in Homa-Bay County, Kenya. The effect of climate change and its variability on crop growth and yields can be determined by its overall effect on soil health and the ability of crops to adjust to shifts in climate and meteorological patterns [10, 37]. While there have been studies on agrometeorological relationship in Homa-Bay including trends in monthly rainfall and temperature in Rachuonyo North subcounty [41] and residents vulnerability to climate change [1], what is missing is the focused study on the effect of climate change on soil security and what this may imply on food insecurity in the area.

Most soil-based studies agree that soil is a product of *in situ* weathering of rock materials or otherwise the accumulation of mineral matter which results to fine earth on the land [38]. Some scholars including [9, 43] even relate soil to a "lifeblood" of vegetation including crops because in their understanding, it provides nutrients, water, and oxygen to plants and support machinery as well as a medium for the decomposition of crops and its residue. For this reason, they emphasize that its utilization and consequently management should ensure land productivity in addition to environmental suitability and sustainability. This therefore requires awareness of soil characteristics which then informs its management procedures. The level of global soil insecurity is almost 2 billion hectares of the total soil resources representing about 22% of the agricultural lands, grazing zones, forest, and other essential vegetation [48, 49]. Usually notable pointers of insecure soil are nutrient depletion such as reduction in soil organic content (SOC) as well as deteriorated soil physical properties such as soil structure, soil texture, and imbalance in soil nutrient status [36]. In Europe, the level of soil insecurity is about 60% to 70%, [5, 15–17]; Asian is about 40% [2], while the insecurity is averagely between 65% and 75% in sub-Saharan Africa [46]. Notably, when on one side increased unsustainable agricultural practices, as occasioned by high food demand by growing human population in Africa, leads to reduction in soil health, it is evident in many studies that Africa's extent of soil insecurity is higher in arid and semiarid lands

(ASALs) as compared to humid zones [7, 29, 45, 50]. The impact of climate change is not limited to atmospheric temperature and change in precipitation patterns but also includes soil temperature [27]. This is expected to have adverse effects on soils and ecosystems [19]. Anthropogenic emissions of greenhouse gases have continued to increase since pre-industrial times, particularly due to global economic and population growth and the resultant cost on climate change and associated self-reinforcement feedback loop especially on soil and food security may be phenomenological [44]. Deforestation, for instance, significantly alters climatic conditions and generates warmer climate hence aridity and desertification [33]. Additionally, while studies including [32] have established the double effect of vegetation to environment—first, emission of carbon (iv) oxide when they are cut down and second, absorption of free atmospheric carbon—the latter implies that forests are natural carbon sinks because they store more carbon through carbon sequestration processes than they release [22]. Further, specialized cropping has also led to increasing use of agricultural inputs such as nitrogen-based fertilizers with the hope of increasing agricultural production [13]. Such fertilizers produce greenhouse gases such as nitrous oxide into the atmosphere due to microbial actions [25]. Regarding extraction of Earth's gravel materials, there is a close relationship between extraction of such earth materials and vegetation cover [6]. This study showed that over the period under investigation, 1985–2022, there was general decline in vegetation cover in areas with intense gravel mining in Narok ward, Kenya, being a case study. It therefore points out to the exposure of the soils in the region to the impact of climate change including increased soil carbon sequestration [32].

Just like many African countries within the sub-Saharan African (SSA) region, Kenya's ecosystems including soils have been adversely affected by climate changes [28, 31]. While this is the case, the importance of her agriculture is by far beyond the limits of domestic consumption, because it also contributes tremendously to the growth of other sectors of the country's economy as a whole [8, 14]. This informs the efforts by Kenya's government to ensure that farming is adequately enhanced with the hopes that this will contribute to the achievement of the famous blue print "Vision 2030," which is anchored on adequate food supply to eliminate food insecurity among the citizens [26].

Accordingly, [35] noted that the soil security situation is affecting the economic livelihood of many people in Kenya. The insecurity aspect does so by conditioning productivity of the cultivable lands which further exhibit a ripple effect on food security [47]. The assessment of the extent of soil conditions in Kenya using remote sensing (RS) and Geographic Information System (GIS) established that most counties in Kenya are facing the risk of soil nutrient depletion and altered physical properties hence highly insecure to support the requisite yield [31]. A case in point and a good example indeed is the situation in Homa-Bay County. The county has great agricultural potential, probably more than any other ASAL counties in Kenya [20, 21]. This potential has never been achieved, yet the county's food security heavily relies on its agricultural production [3, 4, 39] in the face of the county's rapid population growth rate of 2.7% [30].

Being largely rural landscape, the tracks of lands in the county, when cultivated, should produce adequate agricultural yield, mainly maize and sorghum, in addition to others like groundnut, cassava, and sweet potatoes, for subsistence consumption. This is because the area experiences average temperature and moderate rainfall which are perceived conducive for growing these crops [41]. However, most of these parcels of lands, annually, remain uncultivated, while others are also characterized by wide and deep gullies. What is more worrying is the flimsy assurance that the soils in the area are sufficiently secure to support its much needed functions including agricultural activities! The intensified utilization of land in addition to the shift in land-use practices selectively skewed toward non-agricultural practices is also evident. It is worth noting that the absence of sustainable ABPs mostly relating to farming techniques, and by extension the general proper soil utilization, have direct effect on climate change, and this is a threat in itself to soil security. This formed the basis of this study that sought to assess particularly the ABPs causes of climate change and associated cost on soil security and food security in Homa-Bay County. The study aimed at developing pathways for mitigating the effect of climate change on soil and food security in the study area on a sustainable basis.

Misesian's praxeological theory of 1946 was considered relevant to this study. The theory was used to investigate the role of human action or practice on climate change. Mises emphasized on the knowledge to serve human well-being as directly proportional to choices made in the process of endeavoring to achieve self-fulfilment [11, 24]. Human fulfilment requires a set of human endeavor combined with action (praxis), while human endeavor relates to reasoning ability and rational learning to acquire empirical knowledge, actions which are more ethical along with wisdom-based experience [12]. The praxeological "practices" refers to demonstrative knowledge, while this knowledge implies learning as a consequence of action [11, 12]. Mises contextualized praxeological thinking to establish the gravity of human choices in both economic and social sciences [18]. Therefore, human endeavor (practice or action) based on the desire to fulfil the current existing economic needs such as food may be considered biophysical causes of climate change which limits soil security hence likely to cause food insecurity in an area. It is a result of this chain of consequences that provide learning opportunities for human beings to redirect the energy toward adopting alternative and appropriate soil management and conservation strategies.

2 Methodology

Cross-sectional survey design was used to explore the role of anthropogenic-based practices on climate change in Homa-Bay County. It was a mixed study involving both quantitative and qualitative data. Qualitative data were used to examine the opinion of participants about the types and level of anthropogenic-based practices, while quantitative data aided in describing and reporting statistical data on the relationship between the cost of ABPs climate change on soil and food security. The study was conducted in Homa-Bay County (Fig. 1) which has seven sub-counties; Ndhiwa, Rachuonyo North, Rachuonyo South, Suba, Mbita, Rangwe, and Homa-Bay sub-counties. The study area is located within the longitude 34° 45′ E and latitude 0° 45′ S, while it borders Lake Victoria to the west which also extends toward the north, Kisumu and Siaya counties to the north, Migori county to the south, and Kisii and Nyamira counties toward the east. The legend illustrates the types of boundaries and water body present in the map.

The study population was 262,036 households with an average of approximately 5.0 persons per household and population density of 359 persons per unit square kilometer [30]. Further, the population is unevenly distributed in Homa-Bay County. Rachuonyo North was purposively selected from the eight sub-counties in the study area based on the intensity of ABPs, low soil health condition, and high cases of food poverty [20, 21], while a sample size of 250 respondents were also conveniently sampled considering the large size of the study area and the need to minimize the cost of this study. The study used a multistage cluster sampling technique in a manner that in the first stage Homa-Bay County was sampled from the 47

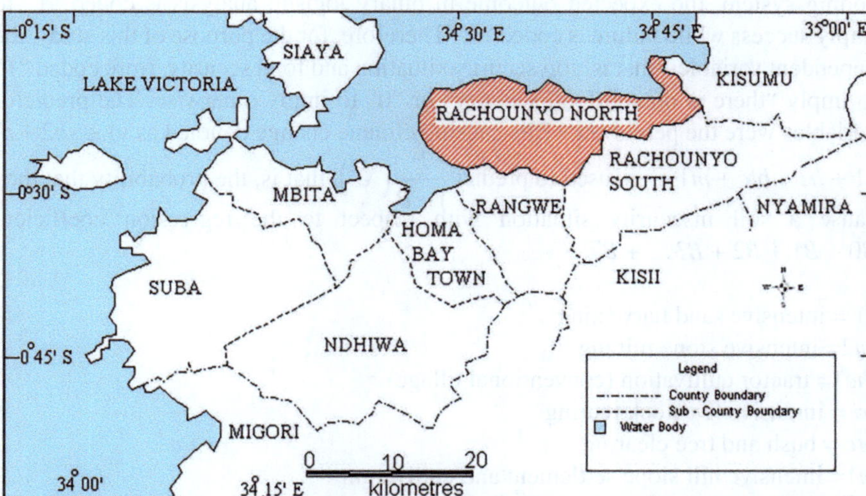

Fig. 1 Map of Homa-Bay County showing administrative units of sub-counties. Source: Own contribution (2024)

counties in Kenya. In the second stage, the eight sub-counties in the study area were considered as eight clusters in which Rachuonyo North sub-county was purposely sampled. Simple random sampling was used in the third stage to draw the samples from each ward of Rachuonyo North sub-county. The data were collected by use of structured questionnaire and observation schedule that were moderated with the help of a co-author. Additionally, 20 key informants were intensively interviewed, and FGD sessions were conducted at ward level to supplement the other data collection tools.

Data analyses involved both descriptive and inferential approaches to statistical data analysis. Descriptive statistical analysis involved calculation of percentages of data on demographic characteristics of the respondents and the level of ABPs that are perceived to cause climate change in the county. Data from the questionnaire were analyzed using SPSS, version 25.0. Significant testing was done using BLRM to establish the relationship between anthropogenic-based practices and soil security at 95% CL. The mathematical representation of the model is illustrated as follows:

$$\frac{P}{1-P}\left(X^*\right) = B0 + B1s1 + B2sn2 + B3thc1 + B4lri + B5btc + B6m1 + B7mc$$

This model describes the probability of an event occurring as a function of X variables in such a manner that, $\frac{P}{1-P}\left(X^*\right)$ is the predicted variable, soil and food security situations, which are dichotomous in nature. In view of [23] binary data coding system, the expected outcome in binary logistic analysis is coded "1" to imply success while failure is coded "0." Therefore, for the purpose of this study, the dependent variables, that is, soil security situation and food security, were coded "1" to imply "there is no soil/food security" or "0" to imply otherwise. The predictor variables were the perceived ABPs causing climate change denoted as $s1 + sn2 + th$ $c1 + lri + btc + m1 + mc$ used to predict $\frac{P}{1-P}\left(X^*\right)$, that is, the probability that they cause a soil insecurity situation with respect to the regression coefficient $B0 + B1 + B2 + B3\ldots + B7$.

$s1$ = intensive sand harvesting
$sn2$= intensive stone mining
$thc1$= tractor cultivation (conventional tillage)
lri = intensive livestock rearing
btc = bush and tree clearing
$m1$= intensive hill slope settlement and cultivation
mc = maize and sorghum cultivation.

3 Results and Discussions

The study considered seven background characteristics of the household heads based on gender, age, length of stay, employment status, and family size, level of monthly income, and education level could help in understanding the variation in responses among the respondents. The findings showed that 50.2% of the respondents were male, while 49.8% were female, and this was an equal gender representation. 54.0% of the respondents were between 40 and 79 years of age, while 29.8% had stayed in the area between 16 and 25 years. Regarding the level of education, the majority, 50.2%, had primary as their highest level of education. 77.2% reportedly had monthly income below Ksh. Ten thousand as unemployment level was at 85.1%. Lastly, 39.1% had family sizes between 7 and 10 members which indicate large family sizes among the households.

4 Common Climate Change-Based Anthropogenic Practices in the Study Area

The findings on the common ABPs related to the effect of climate change in the area showed that 89.1% of the respondents viewed climate change as a consequence of deforestation with the majority of them, 88.4%, agreeing that this had contributed to soil quality reduction. FGDs and KII participants also maintained that cutting of trees is one of the major causes of the shift in climate to harsh conditions in their area. For example, during FGD, one of the participants said the following:

> "Trees are mostly cut for selling as firewood, charcoal burning, building and construction which in my view exposing soils to over-heating."

These findings relate to other studies including [22, 32] that established a linkage between deforestation and vegetation cover such that the absence of adequate vegetation cover, like the case of study area, imposes climate change-related threats such as reduced soil quality conditions which ultimately affect livelihoods. Further, 96.2% of the respondents reported continuous cultivation of crops such as maize and sorghum as common in the study area. These soils are exposed to the effect of climate change with reduction in production viability [13, 25]. Moreover, 71.4% reported poor cultivation methods including maximum tillage particularly the use of tractor and cultivation on the steep hill were also observable, while the majority, 87.9%, of them indicated that farming and settling on the hill slopes subject the soils in these areas to rapid runoffs. Additionally, while 82.3% of the respondents agreed that livestock farming was common in the area, results of focus group discussions show that it entails overstocking and indiscriminate grazing has lowered vegetation cover hence ability to induce climate change threats on soil security. In fact, during one of the FGDs, one of the participants argued as follows:

"We have had challenges with many livestock grazing on our agricultural lands because they clear almost everything, leaving no vegetation behind and that is why personally I can't allow them any chance to graze on my farm."

Concerning stone mining, 78.4% households also reported stone mining as a common practice in their areas of residence, and 80.0% of them reported cases of sand harvesting, and these may cause alteration in vegetation cover [6] which may subsequently lead to climate change [32]. This points out why most of the respondents (92.4%) indicated that stone extraction in the area exposes residents to the impact of climate change. Incidents of stone mining (Table 1) in Rachuonyo North sub-county were highly reported which again varied from ward to another. Additionally, photographs taken (Plate 1) is an evidence of a rural landscape featuring a field with piles of brown stones and esposed soil suggesting an actively ongoing excavation in the area while it also reveals how agricultural lands are progressively turning into quarries together with expansion of derelict lands due to stone mining.

Additionally, the results of BLRA (Table 2) on the relationship between ABP causes of climate change on soil security showed that deforestation (B = 1.177, S.E = 0.530, and $P < 0.026$) was positive and significant, Sand harvesting (B = −0.372, S.E = 0.490, and $P < 0.448$) was negative and statistically insignificant, stone mining (B = 0.906, S.E = 0.661, and $P < 0.170$) was positive and insignificant, livestock farming (B = −1.162, S.E = 1.495, and $P < 0.019$) was negative and statistically significant, monoculture (B = 1.614, S.E = 1.050, and $P < 0.124$) was positive and insignificant, and conventional tillage (B = 1.775, S.E = 1.089, and $P < 0.103$) was positive and insignificant. Lastly, slope-based activities (B = −0.989, S.E = 0.436, and $P < 0.023$) were negative and statistically significant.

These findings therefore show that among the investigated ABPs, deforestation together with live-stocking and slope-based human activities were statistically significant. Therefore, the implored hypothesis that there is no significant statistical relationship between climate change-induced ABPs and soil security in Homa-Bay County was rejected. According to studies including [29, 46, 47, 50], these soils cannot effectively support agricultural production hence exposing the households to food insecurity situations reported at 84.4% (Table 3).

Table 1 Household perception on the level of stone mining in Rachuonyo North sub-county

Wards	No. high	% High	No. low	% Low
Kendu Bay	8	53.3	7	46.7
Kanyaluo	41	82.0	9	18.0
West Karachuonyo	29	72.5	11	27.5
Kibiri	25	83.3	5	16.7
Total/average	**103**	**72.8**	**32**	**27.2**

Source: Own contribution (2024)

Plate 1 Stone mining active on cultivable land in West Karachuonyo ward. Source: Own contribution (2024)

Table 2 Binary logistic regression model

Predictor variables	B	S.E.	Wald	df	Sig.	Exp (B)
Deforestation	1.177	0.530	4.944	1	**0.026***	0.308
Sand harvesting	−0.372	0.490	0.576	1	0.448	0.689
Quarrying	0.906	0.661	1.879	1	0.170	2.474
Live-stocking	−1.162	0.495	5.519	1	**0.019***	0.313
Monoculture	1.614	1.050	2.363	1	0.124	5.023
Conventional tillage	1.775	1.089	2.655	1	0.103	5.901
Slope-based activities	−0.989	0.436	5.149	1	**0.023***	0.372

Model X^2 (7) = 155.57, $p < 0.001$, Pseudo R^2 values Cox and Snell 0.091 Nagelkerke 0.178. n = 250, *Statistically significant
Source: Own contribution (2024)

Table 3 Food security situation in Homa-Bay County

Variables	Frequency	Percentage (%)	Cumulative (%)
There is food security	39	15.6	15.6
There is no food security	211	84.4	100.0
Total	**250**	**100.0**	

Source: Own contribution (2024)

5 Conclusion and Recommendation

This study therefore concludes that while monoculture (96.2%), tractor tillage (71.4%), live-stocking (82.3%), stone mining (92.4%), deforestation (89.1%), and hill slope-based practices (87.9%) are the most common anthropogenic practices inducing climate change in Homa-Bay county, deforestation (B = 1.177, S.E = 0.530, and $P < 0.026$) was positive and significant, livestock farming (B = −1.162, S.E = 1.495, and $P < 0.019$) was negative and statistically significant, and slope-based activities (B = −0.989, S.E = 0.436, and $P < 0.023$) hence are the main cause of soil insecurity in the study area. This explains why most of the households (84.4%) are food insecure.

The study recommends that the County Government of Homa-Bay should establish the Ministry of Forestry and Forest Reserves to enhance forest management services including tree planting and follow-up management of the forests. This will aid in reducing the rate at which soils in the area are lost to wind and water actions.

The County Assembly of Homa-Bay should pass legislations banning settlements and agriculture on the steep slopes of Homa Hills among other high elevated zones as a mechanism toward reducing the surface runoffs, thus controlling the rapid soil transportation on lower elevations.

Further, the study also recommends increased and informed extension services to address issues of poor farming methods which are evident among the household farmers. Such services should be geared toward encouraging maximum retention of crop residues in the farms, avoiding cultivation across the contours, adoption of mixed and rotational cropping, all aimed at maintaining soil security as well as recovering soil nutrients.

The effect of conventional tillage on soil health can be reduced by training households to ensure that learnt tractor-farming skills are applied. This will ensure that there is adequate tillage variation to avoid "hurting" the soil. Additionally, there should be a review of tractor cultivation policies and appropriately encouragement of an alternative minimum tillage such as oxen plough which, though considered primitive, time-consuming, and traditional, is very sustainable.

To address the effect of stone mining on farming lands, the county government through the ministries of land, agriculture, and economic planning, both at the county and national levels, should come up with land-use planning in the county detailing zonation of lands such as zones for settlement, agricultural, and nonagricultural activities including stone and sand harvesting.

Conflict of Interest Researcher declares no conflict of interest.

Data Availability Declaration Data is available in SPSS format and can be shared should there be need.

Funding InformationThis research received no external funding.

References

1. Ajwang, O. J. E. (2017). Assessing the vulnerability of the residents of Rachuonyo north Sub-County to climate variability and climate change. *Journal of Climate, 1*(2).
2. Alam, A. (2014). Soil degradation: A challenge to sustainable agriculture. *International Journal of Scientific Research in Agricultural Sciences, 1*(4), 52.
3. Ambale B. (2018). *Perception on food insecurity and coping strategies among fishing communities in Homa Bay County*. Kenya (Doctoral dissertation, University of Eldoret).
4. Auma, J. O., Lagat, J. K.s., & Nagigi, M. W. (2010). A comparison of male-female household headship and agricultural production in marginal areas of Rachuonyo and Homa Bay District, Kenya. *Jordan Journal of Agricultural Sciences, 6*(4).
5. Bagarello, V. (2017). Effective practices in mitigating soil erosion from field. In *Oxford Research Encyclopedia of Environmental Science*.
6. Barchok, F. K., Mabwoga, S. O., & Kirui, S. (2024). Assessment of land cover change due to stone quarrying in Narok Town Ward. *Journal of the Kenya National Commission for UNESCO, 4*.
7. Bindraban, P. S., et al. (2012). Assessing the impact of soil degradation on food production. *Current Opinion in Environmental Sustainability, 4*(5), 478–488.
8. Birch, I. (2018). *Agricultural productivity in Kenya; barriers and opportunity*. K4D Publishers.
9. Brevik, E. C., & Verheye, W. (2009). Soil health and productivity. In *Soils, plant growth and crop production. Encyclopedia of life support systems (EOLSS), developed under the auspices of the UNESCO*. EOLSS.
10. Brevik, E. C. (2013). The potential impact of climate change on soil properties and processes and corresponding influence on food security. *Agriculture, 3*(3), 398–417.
11. Caldwel, B. J. (1984). Praxeology and its critics; an appraisal. *History of Political Economy, 16*(3), 363–379.
12. Defoer, T. (2000). *Moving methodologies. Learning about integrated soil fertility Management in sub Saharan Africa*. Wageningen University Press.
13. Demirdogen, A., Guldal, H. T., & Sanli, H. (2023). Monoculture, crop rotation policy, and fire. *Ecological Economics, 203*, 107611.
14. Eichsteller, M., Nyagi, T., & Nyukuri, E. (2022). The role of agriculture in poverty escapes in Kenya—Developing a capabilities approach in the context of climate change. *World Development., 149*(2022), 105705.
15. FAO. (2005). *Land and environmental degradation and desertification in Africa*. Food and Agricultural Organization.
16. Feddema, J. J., & Freire, S. (2001). Soil degradation, global warming and climate change impacts. *Climate Research, 17*(2), 209–216.
17. Fischer, G., Nachtergaele, F., Prieler, S., Van Velthuizen, H. T., Verelst, L., & Wiberg, D. (2008). *Global agro-ecological zones assessment for agriculture*. IIASA Publishers.
18. Gasparski, W. W. (1996). Between logic and ethics: The origin of praxeology. *Axiomathes, 7*(3), 385–394.
19. Gelybó, G., Tóth, E., Farkas, C., Horel, Á., Kása, I., & Bakacsi, Z. (2018). Potential impacts of climate change on soil properties. *Agrokémia és Talajtan, 67*(1), 121–141.
20. GoK. (2013). *Homa-Bay county first county integrated development plan (CIDP) 2013–2017*. Government of Kenya.
21. GoK. (2017). *Homa Bay County first county integrated development plan (CIDP) 2017–2022*. Government of Kenya.
22. Gorte, R. W., & Sheikh, P. A. (2010). *Deforestation and climate change*.
23. Hayder, M. H., Young, W., & Sobhani, A. (2016). *A binary logistic regression model of the avoidance maneuvers in two passenger vehicle crasher*. Australia Transport Research.
24. Hieber, D. W. (2017). *Praxeology and language: Social science as the study of human action*. University of California Press.

25. Houser, M., & Stuart, D. (2020). An accelerating treadmill and an overlooked contradiction in industrial agriculture: Climate change and nitrogen fertilizer. *Journal of Agrarian Change, 20*(2), 215–237.
26. Jowi, E. O. (2016). *Is agriculture still the backbone of Kenya's economy?* University of Nairobi, Kenya/Researchgate Publishers.
27. Jungqvist, G., Oni, S. K., Teutschbein, C., & Futter, M. N. (2014). Effect of climate change on soil temperature in Swedish boreal forests. *PLoS One, 9*(4), e93957.
28. Kabubo-Mariara, J., & Kabara, M. (2018). Climate change and food security in Kenya. In *Agricultural adaptation to climate change in Africa* (pp. 55–80). Routledge.
29. Keyzer, P. A., & Sonneveld, M. A. (2011). *The effect of soil degradation on agricultural productivity in Ethiopia: A non-parametric regression analysis*. Researchgate Publishers.
30. KNBS. (2019). *Kenya population and housing census; population of counties and sub-counties* (Vol. 1). Kenya National Bureau of Statistics.
31. LADA. (2016). *Land degradation assessment in Kenya*. Ministry of Environment and Natural Resources.
32. Leon, M., Cornejo, G., Calderón, M., González-Carrión, E., & Florez, H. (2022). Effect of deforestation on climate change: A co-integration and causality approach with time series. *Sustainability, 14*(18), 11303.
33. Longobardi, P., Montenegro, A., Beltrami, H., & Eby, M. (2016). Deforestation induced climate change: Effects of spatial scale. *PLoS One, 11*(4), e0153357.
34. Mishra, B. B., & Roy, R. (2015). Global climate change: Myth, reality and mitigation. Climate dynamics. *Horticultural Science, 2*, 125–139.
35. Muia, V. K., & Ndunda, E. N. (2013). *Evaluating the impact of direct anthropogenic activities on land degradation in arid and semi - arid regions in Kenya*. Kenyatta University Press.
36. Murtaza, G., Ahmed, Z., Usman, M., Tariq, W., Ullah, Z., Shareef, M., et al. (2021). Biochar induced modifications in soil properties and its impacts on crop growth and production. *Journal of Plant Nutrition, 44*(11), 1677–1691.
37. Mutegi, J., Ameru, J., Harawa, R., Kiwia, A., & Njue, A. (2018). Soil health and climate change: Implications for food security in sub-Saharan Africa. *International Journal of Development and Sustainability, 7*(1), 21–33.
38. Nortcliff, S., Hulpke, H., Bannick, C. G., Terytze, K., Knoop, G., Bredemeier, M., & Schulte-Bisping, H. (2000). Soil, 1. In *Definition, function, and utilization of soil*. Ullmann's Encyclopedia of Industrial Chemistry.
39. Nyamusi, N. M. (2017). *Food security, dietary practices and nutritional status of people living with HIV/AIDS in Homa Bay town*. Kenyatta University Institutional Repository.
40. Ojala, M. (2023). Hope and climate-change engagement from a psychological perspective. *Current Opinion in Psychology, 49*, 101514.
41. Opere, E., Juma, S. G., & Sitienei, B. (2016). Trends in monthly rainfall and temperature in Rachuonyo north Sub-County, Kenya. *Ethiopian Journal of Environmental Studies and Management, 9*(6), 738–747.
42. Pill, M. (2022). Towards a funding mechanism for loss and damage from climate change impacts. *Climate Risk Management, 35*, 100391.
43. Schaetzl, R. J., & Thompson, M. L. (2015). *Soils*. Cambridge University Press.
44. Stern, D. I., & Kaufmann, R. K. (2014). Anthropogenic and natural causes of climate change. *Climatic Change, 122*, 257–269.
45. Thiombiano, L. (2007). *Status and trends of land degradation in Africa*. Researchgate Publishers.
46. Tully, K., Sullivan, C., Weil, R., & Sanchez, P. (2015). The state of soil degradation in sub-Saharan Africa: Baselines, trajectories, and solutions. *Sustainability, 7*(6), 6523–6552.

47. Wambua, B. N., & Kithia, S. M. (2014). Effects of soil erosion on sediment dynamics, Food Security and Rural Poverty in Makueni County, Eastern, Kenya. *International Journal of Applied, 4*(1).
48. Xie, H., Yanwei, Z., Zhilong, W., & Tiangui, L. (2020). *A bibliometric analysis on land degradation: Current status, development, and future directions*. Institute of Ecological Civilization, Jiangxi University of Finance and Economics.
49. Ye, L., & Ranst, E. (2009). Production scenarios and effect of soil degradation on long-term food security in China. *Global Environmental Change, 19*(4), 464–481.
50. Zingore, S., Mutegi, J., Agesa, B., Tamene, L., & Kihara, J. (2015). Soil degradation in sub-Saharan Africa and crop production options for soil rehabilitation. *Better Crops, 99*(1), 24–26.

Gender Roles Shifts and Family Cohesion Amidst Climate Change in Wajir West Sub County, Kenya

Nicolette Vanessa Roman, Moses Mutua Mutiso, Hamdi Abdi, and Claire Mutsilila Nakweya

Abstract

Climate change has significantly altered gender roles and disrupted family cohesion in Wajir West Sub-County, Kenya. This study explores how environmental stressors such as droughts and floods have forced women to assume traditionally male responsibilities and how these shifts affect various aspects of family cohesion. The Gender and Development (GAD) framework guided the qualitative research, which involved 16 women from 16 families, purposively sampled for in-depth interviews and focus group discussions (FGDs). Data were analyzed using NVivo software, with strict adherence to ethical protocols, including informed consent and confidentiality. The findings indicate that climate change has strained family cohesion in several key areas. Emotional support within families has weakened, as women, now burdened with increased responsibilities such as securing food, water, and firewood, have less time to nurture familial relationships. Communication between spouses has deteriorated, with many men absent or disengaged, leading to misunderstandings and unresolved conflicts. The parent-child bond has also been affected, as children are often required to assist in household tasks, limiting opportunities for education and leisure activi-

Authors Nicolette Vanessa Roman, Hamdi Abdi and Claire Mutsilila Nakweya contributed equally to this work.

N. V. Roman
Centre for Interdisciplinary Studies of Children, Families and Society (CISCFS), Faculty of Community and Health Sciences, University of the Western Cape, Cape Town, South Africa
e-mail: nroman@uwc.ac.za

M. M. Mutiso (✉) · H. Abdi · C. M. Nakweya
Department of Sociology, Psychology and Anthropology, Moi University, Eldoret, Kenya
e-mail: mosesmutua@mu.ac.ke

N. Ronoh, C. Rono (eds.), *Rethinking Our Ecosystems*, Springer Proceedings in Geography and Sustainability, https://doi.org/10.1007/978-3-031-95781-9_2

ties. Decision-making within households has shifted, with women taking on leadership roles, which has created tensions in marital relationships where traditional power dynamics are challenged. Additionally, women report heightened levels of psychological stress as they attempt to balance caregiving duties with their new economic roles. Instances of gender-based violence (GBV) have risen, particularly when women fail to meet their partners' expectations. While women have formed community-based support networks to mitigate these challenges, these networks are insufficient to fully address the impact on family cohesion. The study concludes that gender-sensitive climate adaptation programs are essential to provide support for women and promote family stability amidst climate change.

Keywords

Climate change · Gender roles · Family cohesion · Wajir West · Kenya · Gender-based violence

1 Introduction

As society confronts the intricacies of climate change, the adaptability of gender roles becomes a vital element in comprehending family dynamics and cohesion. Historical frameworks frequently categorize familial responsibilities, allocating responsibilities and tasks according to conventional gender stereotypes. Recent changes in these positions, driven by environmental challenges and economic need, necessitate a thorough analysis. Families are being compelled to adapt, with individuals transcending traditional boundaries to confront the complex problems presented by climatic catastrophes. This evolution signifies a change in individual identities and impacts collective resilience and adaptability. The examination of these transitions highlights the relationship between evolving gender roles and family cohesion, demonstrating how adaptive tactics can enhance familial relationships under environmental unpredictability. Examining these relationships is essential for formulating comprehensive strategies for climate adaptation and gender equity, so improving societal well-being.

Climate change has become one of the most pressing global issues, disrupting environmental systems and significantly affecting socio-economic structures worldwide. According to the Intergovernmental Panel on Climate Change (IPCC), the frequency and severity of extreme weather events like droughts, floods, and heatwaves have increased due to human activities, with vulnerable communities bearing the brunt of these impacts [1, 2]. As these environmental challenges grow, their socio-economic effects go beyond just ecological harm, leading to major changes in social dynamics, especially regarding gender roles and family cohesion [3, 4].

Though their built infrastructure and capacity to adapt often lowers the immediate consequences, industrialized areas including North America and Europe are nevertheless experiencing socio-economic consequences of climate change [5]. But rural populations in these places are undergoing significant shifts in gender roles as men relocate to cities for work, leaving women to conduct agricultural tasks typically done by men [6, 7]. Men and women negotiating their changing roles in the house disturbs family dynamics and frequently leads to role conflicts and compromises of traditional family unity [8–10]. Particularly in countries like India, Bangladesh, and Vietnam, where agricultural economy thrives, climate change significantly affects rural livelihoods throughout Asia. Apart from upsetting agricultural output, the regular floods in these places force women to assume responsibilities typically occupied by men in order to sustain their families [11, 12]. These developments challenge patriarchal conventions and redefine gender roles in homes, therefore changing the socio-economic scene [13, 14]. Rural communities are also seeing changes in gender roles in Latin American countries including Brazil, Peru, and Colombia as women participate in income-generating activities as a means of adaptation to climate-related agricultural losses [15, 16]. Rising sea levels and strong storms force societies to quickly adjust to new climatic conditions, therefore posing a unique challenge to Oceania especially the Pacific Island states. Women are very important in these fields for local climate adaptation projects; they typically lead campaigns meant to improve community resilience against natural catastrophes [17, 18]. As women increasingly occupy leadership and decision-making roles, conventional gender roles are being disrupted, and family structures and unity are under review as societies try to fit changing climate reality [19, 20]. These developments affect all spheres of community life, not only leadership, thereby redefining gender roles and changing society expectations [21]. Rising temperatures, long-lasting droughts, and altered rainfall patterns cause great effects on Africa, home of some of the most vulnerable populations confronting climate change. Because it relies on pastoralism and rain-fed agriculture, the continent is especially sensitive to climatic changes, which causes food insecurity and economic instability in many different regions [5].

Sub-Saharan Africa clearly shows significant changes in gender roles as women gradually assume financial responsibilities to sustain their family among the collapse in conventional livelihoods [22, 23]. Sometimes this shift in gender dynamics results in disputes inside families when traditional male roles disappear and family cohesion and power dynamics reorganized [24]. East Africa's arid and semiarid soils (ASALs), which make over 80% of the total land area, especially in Kenya suffer greatly from the effects of climate change. Celebrated for their sensitive ecosystems, these regions routinely experience droughts and erratic weather patterns that significantly impact water availability and agricultural output [25]. Women are apparently assuming more and more in these ASAL communities the roles of breadwinners, small business owners, and historically menial resource managers [26–28].

This change causes family relations to change; women are now increasingly involved in household decision-making a responsibility traditionally occupied by men [10, 29]. Because Wajir County, in northeastern Kenya, depends on pastoralism and has limited adaptive capacities, the socio-economic problems are considerably more complex. Mostly from recurrent droughts that destroy cattle populations, which are vital to the local economy, the county suffers major climate-induced disruptions [30]. Many men from the area move to metropolitan areas or nearby counties in pursuit of employment when these conditions worsen, leaving women to take care of homes and assume more financial obligations. This significant change in household dynamics strengthens women by improving their social positions, but it also causes increased family difficulties since conventional expectations are constantly redefining [31]. In Wajir West Sub-County, the significant effects of climate change on the mostly pastoralist livelihoods made the socio-economic changes even more clear. As they adjusted to the negative consequences of climate-induced disturbances, small-scale trading, hand labor, and other occupations, women in Wajir West became the main breadwinners.

This study was necessary to the great influence these changing gender roles had on Wajir West family cohesiveness. The disturbances brought about by climate change not only changed the social roles that homes occupied but also questioned the cultural standards and expectations that had been ingrained about gender. The stability of families suffers greatly from this redefining since conventional family structures have changed and new chances for empowerment as well as fresh causes of conflict have evolved. Awareness of these dynamics is essential to create gender-sensitive adaptation strategies that can help men and women manage the socio-economic changes brought about by climate change [32].

This study aimed to fill a critical gap in the literature by examining how gender role shifts, driven by climate-induced stressors, impacted family cohesion in Wajir West. While much of the existing research had focused on the economic aspects of climate change, there was a pressing need to understand the socio-cultural dimensions, particularly in regions like Wajir West, where traditional gender roles were undergoing rapid transformation [33]. By investigating the lived experiences of women and the broader community dynamics in this context, the study sought to provide insights into how policies could be designed to not only empower women but also strengthen family cohesion in the face of climate adversity.

The findings of this study had the potential to inform policymakers, development practitioners, and community leaders on the importance of integrating gender-sensitive approaches into climate adaptation strategies. As Wajir West continued to grapple with the effects of climate change, fostering inclusive strategies that considered the evolving roles of both genders was vital to achieving sustainable development and enhancing social resilience [34]. Addressing these challenges holistically not only contributed to better adaptation outcomes but also promoted equitable growth and stability within households and communities in Wajir West and beyond.

2 Theoretical Framework

The Gender and Development (GAD) framework, developed in the late 1970s and early 1980s as a critique of Women in Development (WID), underpins this study. In the 1980s, feminist researchers and practitioners like Caroline Moser helped advance the GAD approach [35]. GAD examined the social construction of gender roles and power relations that shaped gender dynamics in socio-economic and political situations. Its basic idea was that gender roles were socially constructed and evolved across cultures and periods, which had important implications for analyzing how development policies and practices influenced them [36]. Gender connections and how social norms and power structures affected men and women were fundamental to the GAD framework. This perspective allows for a full investigation of gender inequality and their effects on men and women. Intersectionality, which recognized that race, class, ethnicity, age, and gender interacted to create various experiences and challenges, was also part of the GAD approach [37]. This inclusive perspective helped explain gender inequality's complexity and development effects. The GAD framework emphasized empowerment and active engagement, which was its strength. GAD sought radical changes to address gender inequality, unlike WID, which sought to include women into development frameworks without questioning fundamental systems. Women's participation in public and home decision-making promoted gender-sensitive policies that attempted to undermine patriarchal structures and promote equitable development [38]. To comprehend gender issues holistically, the GAD framework included social, cultural, and political factors beyond economic measures. Although strong, the GAD framework had drawbacks. Implementing the framework in conservative or patriarchal countries where structural changes were resisted made it difficult to achieve its transformative goals [39]. The framework's focus on intersectionality and multiple social variables required a nuanced understanding of overlapping factors that may not be easily measurable or addressable in practice, which could complicate policy design and analysis [40]. As comprehensive methods to alleviate deep-seated disparities required major financial and institutional investments, resource constraints in low-income countries hindered GAD's proposals. GAD provided a relevant and insightful lens for this study on gender roles and family cohesion in Wajir West, Kenya, during climate change. The study revealed that climate change forced women in Wajir South to play male duties, disrupting gender norms and family power relations. The GAD framework's focus on gender relations and empowerment allowed researchers to examine how these changing roles influenced women's position in the household and the community [41]. This study used the GAD method to emphasize the need for gender-sensitive adaptation measures that empowered women and promoted family cohesiveness, reflecting the framework's commitment to inclusive development. Thus, the Gender and Development framework, which seeks to change gender relations, empower women, and address structural inequities, provided a solid theoretical foundation for our study. It fits the study's goals of understanding the sociocultural and economic effects of climate-induced gender role shifts in Wajir South, providing a theoretical foundation that supported gender dynamics analysis and informed inclusive adaptation strategies to climate change [14].

3 Methodology

This study aimed to explore the impact of climate-induced shifts in gender roles on family cohesion in Wajir West Sub-County, Kenya, using a qualitative approach. The study was chosen due to its arid and semiarid climate, which has been significantly impacted by climate variability. The target population consisted of women from households in Wajir West, who play a critical role in responding to the socio-economic challenges posed by climate change. The final sample consisted of 16 women selected to ensure that participants had direct experience with the shifts in gender roles and family dynamics.

Data collection was conducted using both in-depth interviews and focus group discussions (FGDs). The 16 women participated in one-on-one in-depth interviews, while two focus group discussions were organized, each comprising of eight participants. These groups were deliberately mixed to include a variety of women based on their age, marital status, and economic roles, facilitating dynamic interactions and the exchange of diverse viewpoints. The FGDs provided a platform for women to collectively discuss their experiences and the broader social implications of changing gender roles on family cohesion.

Thematic analysis was conducted using thematic analysis, a method well-suited for identifying and interpreting patterns within qualitative information. All audio recordings from the interviews and focus group discussions were transcribed verbatim, and a coding framework was developed to categorize the data into key themes related to gender role shifts, socio-economic changes, and their impact on family cohesion. NVivo software was utilized to organize and manage the data, allowing for a systematic approach to identify recurring themes and relationships within the narratives shared by the participants.

Ethical considerations were upheld throughout the study to protect the rights and dignity of the participants. All women involved in the study were fully informed about the purpose of the research, the procedures involved, and their rights as participants. Informed consent was obtained from each participant, with assurances that their information would remain confidential and that their participation was entirely voluntary. All data collected were anonymized to maintain the privacy of the respondents and respect the sensitive nature of the information provided.

This robust methodological approach ensured that the study was conducted with the highest ethical standards while employing rigorous qualitative techniques to gather and analyze data. By focusing exclusively on women in Wajir West Sub-County and incorporating both individual and group perspectives, the study provided a comprehensive understanding of how climate change has reshaped gender roles and family cohesion.

4 Results and Discussions

4.1 Profile of Respondents

In this section, the demographic characteristics of the respondents involved in the study on "Gender Roles Shifts and Family Cohesion amidst Climate Change in Wajir West Sub-County, Kenya" are presented and analyzed. The information provides a clearer understanding of the social and economic backgrounds of the participants, allowing for more nuanced interpretations of their responses and the overall findings of the study. The demographic data include variables such as age, marital status, type of family, number of children, ethnicity, level of education, religion, and occupation. These characteristics are summarized in Table 1.

The demographic characteristics of participants, summarized in Table 1, highlight the interconnected roles of age, marital status, and cultural norms in shaping household dynamics amid climate stressors in Wajir West. Half of the respondents (50%) were aged 21–42, with 37.5% between 43 and 63 years, reflecting a concentration of individuals in their prime working and caregiving years. This aligns with global trends where middle-aged adults, particularly women in rural settings, shoulder heightened responsibilities during socio-economic disruptions [42, 43]. In Wajir West, this demographic face acute pressures as climate-induced male migration forces women into dual roles as providers and caregivers, straining traditional gender expectations.

Marital status further delineates vulnerabilities: 37.5% of participants were married, 37.5% single, 18.8% widowed, and 6.2% divorced. The high proportion of widowed women underscores the precariousness of female-headed households in contexts lacking social safety nets. Similar patterns are observed in Nepal, where widows navigating climate shocks report intensified economic and social isolation [44, 45]. Divorced women in Wajir West similarly struggle to reconcile caregiving with income generation, emphasizing how marital instability exacerbates gendered burdens. These findings call for targeted interventions, such as subsidized childcare and widow-focused livelihood programs, to mitigate compounding stressors.

Family structure emerged as a critical factor, with 56.2% of respondents in polygamous householdsa norm rooted in Somali pastoralist traditions. While polygamy historically distributes labor, climate-driven resource scarcity strains cooperation among spouses, amplifying competition for limited resources. Recent studies in sub-Saharan Africa corroborate this, noting that polygamous households experience heightened economic fragility during droughts compared to nuclear families [46]. Nuclear households (43.8%), though smaller, face parallel challenges, as shifting gender roles disrupt decision-making hierarchies. This duality underscores the need for adaptive policies, such as cooperative resource-sharing models, tailored to diverse family systems.

Table 1 Demographic characteristics of respondents

Variable	Category	Frequency	Percentage (%)
Age of respondents	≤ 20	1	6.2
	21–42	8	50
	43–63	6	37.5
	64+	1	6.2
Total		16	100
Marital status	Divorced	1	6.2
	Married	6	37.5
	Single	6	37.5
	Widowed	3	18.8
Total		16	100
Type of family	Nuclear	7	43.8
	Polygamous	9	56.2
Total		16	100
Have children	Yes	11	68.8
	No	5	31.2
Total		16	100
Number of children	≤ 0	5	31.2
	4-7	4	25
	8-9	5	31.2
	9+	2	12.5
Total		16	100
Ethnicity	Somali	16	100
Total		16	100
Level of education	None	4	25
	Primary	3	18.8
	Secondary	6	37.5
	Tertiary	3	18.8
Total		16	100
Religion	Muslim	16	100
Total		16	100
Occupation	Housewife	5	31.2
	Self-employed	6	37.5
	Unemployed	5	31.2
Total		16	100

The presence of children intensified pressures: 68.8% of respondents had children, with 31.2% managing five or more dependents. Larger families (5–8 children: 31.2%; 9+ children: 12.5%) reported acute financial strain, mirroring findings in Latin America where high dependency ratios correlate with maternal stress and intra-household conflict [47]. Even smaller households grappled with balancing childcare and income generation, reflecting rigid gender norms that prioritize women's domestic roles. These challenges highlight the urgency of community-based childcare initiatives to reduce caregiving burdens and enable women's economic participation.

Ethnic and religious homogeneity further shaped responses. All participants identified as Somali Muslims, a demographic uniformity reinforcing patriarchal norms resistant to role shifts. As observed in Tanzanian Muslim communities, such cultural rigidity often stigmatizes deviations from traditional gender roles [21]. In Wajir West, this resistance intersects with low educational attainment, 25% had no formal education, and only 18.8% attained tertiary levels to constrain women's access to stable livelihoods. Recent studies emphasize education as a buffer against economic shocks [48, 49], yet limited schooling confines many women to informal, precarious work (37.5% self-employed; 31.2% unemployed). Addressing these barriers requires adult education programs and partnerships with religious leaders to align gender equity with cultural values.

5 Emotional Support Within the Household

Emotional support within families entails the provision of care, empathy, and mutual understanding among members, acting as a vital foundation for household stability and unity, particularly during crises. In Wajir West Sub-County, climate change has catalyzed significant shifts in traditional gender roles, profoundly altering these emotional support systems. As recurrent droughts and floods compel women to juggle income-generating activities and domestic duties, their capacity to nurture emotional bonds within families diminishes. This strain manifests in heightened spousal conflicts, emotional disengagement from children, reduced intimacy between partners, and the erosion of extended familial networks. Table 2 delineates

Table 2 Emotional support within the household

Main theme	Sub-theme	Key findings
Emotional support	Strain on spousal relationships	Women taking on primary breadwinner roles led to emotional disconnection
		Role reversals caused tension and conflict in marriages
		Spouses struggled to communicate due to stress and frustration
	Emotional detachment from children	Mothers experienced difficulty in balancing caregiving and economic roles
		Reduced emotional interactions between mothers and children
		Increased feelings of guilt among women for not being emotionally available for their children
	Reduced spousal affection	Economic hardship reduced emotional intimacy between spouses
		Conversations between spouses shifted from affection to survival-focused interactions
		Spouses felt emotionally distant, weakening marital bonds
	Impact on extended family support	Climate stress weakened extended family support systems
		Families became more isolated as extended networks focused on individual survival
		Traditional forms of emotional and practical family support declined

the sub-themes emerging from this study, illustrating how climate-driven role reconfigurations disrupt emotional support mechanisms.

The realities of climate change in Wajir West Sub-County have profoundly reshaped gender roles and family cohesion, creating new social and emotional challenges within households. As extreme weather events such as droughts and floods disrupt traditional livelihoods, the redistribution of responsibilities has triggered significant shifts in household dynamics. These transformations, driven primarily by economic survival imperatives, have redefined spousal relationships, strained parent-child connections, weakened marital intimacy, and eroded the extended family support system. The findings, as outlined in Table 2, illustrate how these shifts are altering family structures, often in ways that intensify vulnerability and emotional distress.

One of the most visible consequences of these gender role shifts is the growing strain on spousal relationships. Traditionally, men in Wajir West Sub-County have been the primary providers, ensuring financial security for their families. However, as prolonged droughts and floods decimate pastoral livelihoods, many men have migrated to urban areas in search of alternative employment, leaving women to assume both economic and caregiving responsibilities. This reversal of roles has disrupted the conventional balance of power within households, creating tensions and redefining marital dynamics. Women who once relied on their husbands for financial stability now find themselves as primary breadwinners, while men struggle with feelings of displacement and loss of identity. A participant shared her frustration: "My husband used to be the one we relied on, but since the droughts hit, he left to find work, and now I'm the breadwinner. When he calls, all we do is argue because he thinks I've taken over his role" (Participant 3, individual interview). Another respondent noted: "The floods made things worse. My husband is never home anymore, and when he does come back, he's too angry to talk. It feels like we're drifting apart" (Participant 7, focus group discussion).

This pattern of role reversal leading to spousal tension has been observed in other climate-vulnerable regions. Research in Bangladesh and South Africa confirms that climate-induced male migration disrupts traditional gender roles, often resulting in increased stress and conflict within marriages [50, 51]. Similar findings in Kenya and Ethiopia indicate that men who return home after prolonged absences often feel alienated from household decision-making, exacerbating marital strain [10, 52]. The emotional impact of these changes is not only felt by men struggling with their altered roles but also by women, who, while gaining economic responsibilities, experience increased exhaustion and emotional detachment from their spouses. A participant reflected on this: "We used to make decisions together, but now I do everything on my own. When he comes back, he feels left out, and this causes a lot of tension between us" (Participant 10, individual interview). The growing dissonance between men and women in these contexts underscores the urgent need for targeted interventions that promote role adaptation strategies and provide psychosocial support to families navigating these shifts. Without such interventions, the risk of deteriorating marital bonds and family breakdowns will continue to rise.

Shifting economic responsibilities have also affected parent-child interactions, particularly between mothers and their children. As women assume more economic duties, their ability to provide emotional and social support has declined, leading to a growing sense of detachment between them and their children. A mother in this study described her experience: "I'm always on the move, trying to find money to feed my kids. I don't have time to sit with them or listen to their stories. They look at me like I'm a stranger, and it hurts so much that I can't be the mother I used to be" (Participant 13, individual interview). Another participant reflected on how financial stress affects her ability to engage with her children: "Even when I'm home, my mind is somewhere else, worrying about how to get through another day. They say I'm always angry, but I'm just scared and exhausted" (Participant 8, focus group discussion).

These experiences align with findings from Ethiopia and India, where research indicates that when women's financial responsibilities increase due to climate-induced stressors and maternal engagement decreases, negatively impacting children's emotional and social well-being [53–55]. The lack of maternal presence contributes to behavioral challenges, poor academic performance, and an increased sense of neglect among children. In West Africa, similar patterns have been observed, where climate-induced economic strain reduces women's availability for caregiving, leading to weakened parent-child relationships and heightened risks of childhood emotional distress [56, 57]. These findings suggest that climate-induced economic hardships not only disrupt gender roles but also have intergenerational consequences, underscoring the importance of support systems that allow mothers to balance financial responsibilities without compromising their emotional availability to their children. Without such support, the long-term development of children in climate-affected regions may be severely compromised.

The emotional strain caused by climate-induced economic instability has also led to a marked decline in spousal affection. As financial stress dominates daily interactions, conversations have shifted from expressions of love and connection to discussions of survival and loss. One respondent captured this shift: "We used to talk and laugh even when times were tough, but now all our conversations are about what we've lost. I miss those moments, but all I have left is stress and no energy to be loving or kind" (Participant 5, individual interview). Another participant described how the pressures of economic survival have eroded intimacy in her marriage: "My husband and I don't even look at each other the same way anymore. The drought took his job and my peace, and now we're just trying to survive under the same roof without hurting each other more" (Participant 6, focus group discussion).

These findings align with broader research, which has shown that economic hardship and displacement caused by climate change contribute to a decline in spousal intimacy. A study in Mali confirmed that as financial pressures increase, marital relationships become more transactional and less affectionate [58, 59]. Research in the aftermath of natural disasters in the USA and Bangladesh similarly found that financial distress and post-disaster trauma contribute to higher rates of marital dissatisfaction and domestic violence [60–62]. These findings emphasize the need for psychosocial and mental health interventions that support couples in sustaining emotional connections despite external pressures.

Compounding these household challenges is the erosion of extended family support systems. Traditionally, extended families in Somali communities played a crucial role in providing financial and emotional assistance during times of crisis. However, climate-induced hardships have strained these networks, as many relatives now struggle to support themselves. A respondent in this study shared: "I used to rely on my sisters and cousins when things got rough, but now we're all struggling. No one has time for emotional support because everyone's trying to survive" (Participant 11, individual interview). Another participant noted: "The floods have made us so isolated. We can't support each other like we used to because we're all fighting our own battles" (Participant 14, focus group discussion).

This weakening of extended family support structures has been observed in climate-affected regions across Africa. Studies in West Africa and Bangladesh show that climate pressures are fragmenting kinship structures, reducing traditional forms of mutual assistance and increasing social isolation [63–66]. As families become more economically vulnerable, their ability to support one another declines, reinforcing cycles of hardship. This underscores the need for community-based initiatives aimed at rebuilding social cohesion and strengthening extended family networks in the face of climate-induced distress.

As gender roles continue to shift in response to climate change, the impact on family cohesion in Wajir West Sub-County remains profound. Spousal relationships, parent-child bonds, marital intimacy, and extended family support structures are all undergoing significant transformations, often at the cost of emotional well-being. Addressing these challenges requires a comprehensive approach that integrates economic empowerment, psychosocial support, and community resilience strategies. Climate-sensitive gender policies and social protection programs can mitigate these disruptions, enabling families to adapt while maintaining their social and emotional stability. Understanding the intersection of climate change, gender, and family dynamics is crucial for designing interventions that foster resilience and well-being in vulnerable communities.

6 Communication

Communication is a fundamental process within households that serves as a vehicle for emotional connection, decision-making, and managing everyday tasks. In the context of gender role shifts caused by climate change, communication patterns within families often undergo significant transformations. Climate change-induced stressors, such as droughts and floods, have forced many men to migrate in search of economic opportunities, leaving women to manage both the household and income-generating activities. This shift has profoundly affected the way family members communicate, altering the dynamics of decision-making and emotional support within families. Effective communication is essential in maintaining family cohesion and adapting to new responsibilities. However, when traditional gender

Table 3 Communication in the context of gender role shifts and climate change

Main theme	Sub-theme	Key findings linked to gender role shifts and climate change
Communication	Breakdown in spousal communication	Gender role shifts, where women become primary breadwinners due to men migrating for work, have led to miscommunication and tension within marriages. Spouses argue more as roles shift, particularly around decision-making and economic pressures, with men feeling displaced and women overburdened
	Survival-centered communication	Climate change stressors, such as droughts and resource scarcity, have shifted communication toward survival-focused discussions (food, water, shelter) rather than emotional or future-oriented conversations. This survival-centric focus often reduces emotional connection
	Communication gaps with children	As women take on dual roles both managing the household and earning income their reduced time at home has resulted in limited communication with their children, leading to emotional gaps and weakened family cohesion. Climate change has intensified the need for these economic roles, exacerbating the issue
	Lack of collaborative decision-making	With gender roles shifting, women have assumed decision-making authority in many households, while men, often absent due to migration, feel excluded from important decisions. This dynamic increase tension and reduces collaborative efforts within the family
	Reduced communication with extended family	Climate-induced economic hardships have weakened communication with extended family members, as households focus on immediate survival. This erosion of communication weakens traditional extended family support systems, which were once integral in providing collective decision-making and support

roles are disrupted, communication can become strained, leading to misunderstandings, emotional distance, and weakened familial bonds. The study revealed several communication-related issues linked to gender role shifts and climate change. These challenges are summarized in Table 3.

The findings in Table 3: Communication in the context of gender role shifts and climate change illustrate how climate-induced changes in gender roles have significantly impacted family communication in Wajir West Sub-County. Traditionally, men served as the primary breadwinners, while women focused on caregiving and household management. However, climate change particularly prolonged droughts, and resource scarcity has forced many men to migrate for work, leaving women to assume both economic and domestic responsibilities. This shift has redefined household interactions, creating tensions in spousal relationships, survival-driven communication, weakened parent-child connections, reduced collaborative decision-making, and diminished ties with extended family members.

One of the most pressing issues is the breakdown in spousal communication due to shifting responsibilities. As women become the primary providers, many men feel displaced, while women struggle with the burden of dual roles. The change has altered how couples interact, often leading to frustration and conflict over decision-making and financial pressures. Several participants highlighted this in focus group discussions. One woman explained, "Now that I make most of the decisions, conversations with my husband have changed. We no longer talk about our future, just about what is missing, and it often ends in arguments" (Participant 6, FGD). Another shared, "My husband used to take charge, but now he speaks as if he resents me for taking over. We don't talk like before; it's all frustration" (Participant 10, individual interview). Similarly, a third participant noted, "He barely speaks when he comes home. He's angry that he's not the provider anymore, and I'm managing everything. It's not just my family; many others are going through this" (Participant 3, FGD).

These findings align with studies from Bangladesh and Ghana, where climate-induced economic shifts have led to similar marital tensions. Research indicates that as women assume financial responsibilities, men often withdraw emotionally, resulting in reduced communication and relationship strain [67, 68]. In Ethiopia, studies show that men's migration and economic instability contribute to emotional disconnection within marriages, further weakening family cohesion [69]. Without effective communication, relationships deteriorate, increasing the risk of conflict and separation. Addressing this requires interventions such as couple counseling and programs that support men in adjusting to evolving gender roles, ensuring that marital bonds remain strong despite shifting responsibilities.

Another emerging trend is the shift toward survival-centered communication in households. Previously, families engaged in discussions about long-term goals, education, and household improvements. However, climate stressors have pushed conversations toward immediate survival concerns such as securing food, water, and shelter rather than emotional or future-oriented topics. One woman described this shift, stating, "We don't have time to talk about anything else. Every conversation is about how to get through the day whether we'll have food or enough water" (Participant 2, individual interview). Another participant emphasized the communal nature of this shift, explaining, "Everyone in the community talks about the same thing struggling for food, firewood, and water. There's no time for anything else" (Participant 7, FGD).

This phenomenon has been observed in other climate-vulnerable regions. In Bangladesh, research found that under severe climate stress, families prioritize daily survival over emotional bonding, resulting in a loss of meaningful interpersonal connections [70]. Similarly, studies in West Africa show that climate-induced economic hardship forces families into crisis management mode, leaving little room for nurturing relationships [71]. The long-term impact of this shift is significant. The lack of future-oriented discussions erodes family cohesion, making relationships transactional rather than supportive. Additionally, children's psychological well-being is at risk, as the absence of emotional engagement deprives them of the security and stability needed for healthy development. Addressing this requires economic interventions to ease household burdens and mental health support programs that help families rebuild emotional connections despite ongoing survival pressures.

The study also highlights a growing communication gap between parents and children. As women take on both caregiving and financial responsibilities, their availability for their children diminishes. Many mothers reported feeling emotionally distant from their children, as their primary focus has shifted to providing for basic needs rather than engaging in meaningful interactions. One participant expressed, "I barely have time to sit with my children anymore. My mind is always on how to get the next meal" (Participant 4, Individual Interview).

Similar findings have been documented in West Africa and India, where research indicates that as women take on greater economic roles, communication with their children declines, creating emotional gaps and weakening family bonds [8, 23]. Studies suggest that when parental communication is limited, children experience lower emotional resilience, behavioral issues, and diminished aspirations [72]. Social support programs should be introduced to help women balance their responsibilities, ensuring that parent-child relationships remain intact despite economic hardships.

Another consequence of climate-induced gender role shifts is the erosion of collaborative decision-making within households. With men frequently away for work, women increasingly make household decisions alone. However, when men return, they often feel excluded from family affairs, leading to tension and disputes. One respondent shared, "When my husband comes back, he feels like he has no say in anything anymore, and this leads to arguments. He says I've taken over, but I had no choice" (Participant 8, individual interview).

This pattern aligns with findings from Ghana and Mali, where research indicates that as women assume greater household authority due to men's migration, traditional decision-making structures are disrupted, increasing spousal conflicts [73, 74]. The breakdown of joint decision-making weakens trust and cooperation within marriages, making it essential to develop strategies that allow both partners to remain involved, even when separated. Community programs that encourage inclusive decision-making and provide remote communication channels for migrating men could help mitigate this challenge.

The study further highlights the decline in communication with extended family members, traditionally an essential part of Somali community resilience. Economic hardships caused by climate change have forced families to focus solely on their immediate survival, reducing their ability to maintain ties with extended relatives. One respondent explained, "We used to rely on extended families for support, but now everyone is struggling. We don't talk like we used to" (Participant 11, individual interview).

This trend mirrors findings from West Africa, where climate stress has led to the fragmentation of kinship networks, weakening traditional social safety nets and increasing household isolation [75]. The loss of extended family support places additional stress on nuclear families, increasing their vulnerability to financial and emotional strain. Strengthening community-based initiatives that encourage extended family engagement could help rebuild these support networks and enhance household resilience.

The findings in Table 3 highlight how climate change has redefined family communication in Wajir West Sub-County, reshaping spousal dynamics, parent-child interactions, and broader social relationships. As traditional roles evolve, tensions emerge within marriages, survival-focused communication replaces emotional bonding, and extended family connections weaken. Addressing these challenges requires a holistic approach, integrating economic relief measures, psychosocial support services, and gender-sensitive programs that help families navigate these changes while maintaining cohesion. Without such interventions, the social fabric of climate-affected communities will continue to erode, with lasting implications for family well-being and resilience.

7 Financial Conflicts and Household Tensions

Financial disputes have become a major source of conflict in households as women increasingly take on financial decision-making roles. With men frequently absent or unemployed due to climate-induced economic disruptions, many families now rely on women for income. However, men often struggle to relinquish financial control, leading to power struggles and mistrust.

One participant shared, "Handling the money has turned into a war at home. My husband accuses me of spending too much, but he doesn't understand that every coin goes to feeding the family" (Participant 12, individual interview). Another woman explained, "The floods left us broke, and now every time I try to buy something essential, my husband sees it as a waste" (Participant 15, focus group discussion).

Similar challenges have been reported in Ethiopia, where women assuming financial responsibilities due to male migration led to increased marital conflict [69]. In Bangladesh, financial disputes intensified when women became primary earners in response to climate-related livelihood disruptions [76]. Addressing these conflicts requires financial literacy programs and joint decision-making frameworks to foster trust and cooperation within households.

8 Declining Joint Problem-Solving in Households

The study reveals a decline in joint decision-making, as climate-induced economic pressures force women to assume leadership roles in managing family affairs. Previously, spouses collaborated on household decisions, but as men disengage, women increasingly bear the burden alone.

One woman explained, "We used to make decisions together, especially on big issues, but now it's just me. I feel like I'm carrying the weight alone" (Participant 4, individual interview). Another added, "Since the droughts, it's like everyone has their own problems to solve, and we're not doing it together anymore" (Participant 3, focus group discussion).

This shift weakens family cooperation and increases emotional distance between partners. Similar trends have been observed in Nepal, where environmental stressors and male migration led to women assuming full household responsibility, resulting in heightened stress and reduced collaboration [77]. In Senegal, men returning from labor migration often struggled to reintegrate into household decision-making, further straining family cohesion [78]. Strengthening family dialogue, providing training on joint decision-making, and fostering inclusive leadership approaches can help rebuild cooperation.

9 Gender-Based Violence and Family Cohesion

As women take on economic and leadership responsibilities, shifting gender roles have triggered increased household tensions, sometimes escalating into gender-based violence (GBV). These conflicts manifest in different forms, including physical violence, psychological abuse, economic coercion, and social isolation, all of which threaten family stability.

10 Increased Physical Violence

Findings indicate that some men react with aggression when they feel their traditional authority is being undermined by women's new roles.

One woman explained, "Ever since I took over managing the finances because of the drought, my husband has become more aggressive. He thinks I'm taking his place" (Participant 3, individual interview). Another added, "The floods have made everything harder. My husband gets violent whenever he feels like he's losing control over our household decisions" (Participant 5, focus group discussion).

Similar trends have been reported in Uganda, where economic hardship caused by climate change has led to an increase in domestic violence as men struggle to adapt to shifting gender roles [9]. In Nigeria, financial stress due to droughts has been linked to a rise in intimate partner violence [79]. Addressing these challenges requires mental health support, conflict resolution programs, and awareness campaigns to promote gender-equitable household transitions.

11 Psychological and Economic Abuse

Beyond physical violence, psychological and economic abuse have emerged as additional challenges, as men seek to retain financial dominance despite contributing less to household income. Women reported experiencing emotional manipulation, blame, and restrictions over financial control.

One woman shared, "Even though I'm earning the money now because of the droughts, my husband still controls how we spend it. He doesn't trust me to make the right decisions" (Participant 8, individual interview). Another added, "The verbal abuse never stops. He keeps saying it's my fault that we're in this mess, even though it's the floods that ruined everything" (Participant 7, focus group discussion).

Similar findings have been documented in Malawi, where increased economic responsibility among women due to climate stressors led to financial coercion and psychological distress [80, 81]. In Cambodia, climate-driven household economic changes have been linked to an increase in controlling behaviors among men [82]. Implementing community-based interventions, financial empowerment programs, and counseling services can help women navigate these challenges while improving family harmony.

12 Social Isolation and Gender Role Shifts

One significant consequence of climate-induced gender role changes in Wajir West Sub-County is the social isolation of women. As they take on greater financial and decision-making responsibilities, some men respond by limiting their mobility and interactions to reinforce traditional gender hierarchies. This isolation weakens family cohesion, restricting women's access to support systems needed to navigate their evolving roles.

A participant shared, "He doesn't allow me to meet with other women in the community anymore, saying they fill my head with ideas that make me disrespect him" (Participant 9, individual interview). Another added, "Since the floods, the more I try to be independent, the more he isolates me from my friends and family" (Participant 12, focus group discussion).

Similar trends have been observed globally. Studies in India [83] and Ethiopia [84] found that climate-induced economic shifts led to men restricting women's interactions to limit their growing autonomy. This not only affects women's psychological well-being but also reduces communication within households, increasing family tensions.

Interventions that engage men in adaptation strategies, strengthen women's networks, and create safe spaces for discussion can mitigate these effects. Research in Bangladesh [85] found that when women maintain access to social and economic opportunities, household tensions decrease, and climate resilience improves.

13 Cultural Resistance and Changing Perceptions

Despite women's expanding roles in household management and decision-making, resistance to these shifts remains strong. Many still associate leadership and financial provision with men, creating friction within families as economic realities shift.

One participant explained, "Many people still expect me to stay at home despite the drought forcing me to work outside" (Participant 11, individual interview). Another added, "Men feel threatened by women taking on leadership roles, seeing it as a challenge to their authority" (Participant 13, focus group discussion).

This resistance mirrors findings from West Africa [86] and Pakistan [87], where climate-driven gender role changes led to household disputes and social stigma against women taking leadership roles.

However, perceptions are gradually shifting, as communities recognize women's contributions to household survival. One woman noted, "Some now see the value of what we do. Without our efforts during the droughts, many families wouldn't have survived" (Participant 14, individual nterview). Similar shifts were documented in Rwanda [88], where women-led climate adaptation efforts increased gender inclusivity. Encouraging education programs and community discussions can further reduce resistance and support household stability.

14 The Role of Traditional Leaders in Household Adaptation

Traditional leaders play a key role in shaping gender norms in Wajir West Sub-County, influencing whether families resist or accept women's expanded roles. Encouragingly, some elders are beginning to support these changes as they witness the necessity of women's leadership.

One participant shared, "The elders are starting to support us because they see that women are essential in making tough decisions during the droughts" (Participant 16, individual interview).

Research in Ethiopia [89] and Malawi [90] found that when traditional leaders endorse women's leadership, family adaptation improves, and household conflicts decrease. Engaging elders in climate resilience and gender inclusion efforts can help normalize these transitions.

15 Women's Leadership and Family Resilience

As climate change reshapes household structures, women's leadership is emerging as a crucial factor in family resilience. Many women have stepped into leadership roles out of necessity, ensuring household survival during crises.

One participant explained, "I never imagined leading others, but now I'm guiding women in our community to adapt and support their families" (Participant 4, individual interview).

Studies in South Asia [91] and Latin America [92] confirm that women's leadership in climate adaptation strengthens household stability and economic security. To sustain this progress, investments in leadership training, financial inclusion, and legal protections are essential.

16 Limitations of the Study

The study encountered several limitations that were carefully considered and addressed where possible. One primary limitation was the reliance on qualitative data, which, while rich in context, may not be entirely generalizable to the wider population of Wajir West Sub-County. To mitigate this, the study employed diverse sampling techniques, ensuring participants were selected from various socio-economic backgrounds and demographic groups to capture a broader spectrum of perspectives on gender role shifts and family cohesion.

Another limitation was the potential for response bias, particularly in discussing sensitive issues like gender dynamics and family relationships. Participants might have felt uncomfortable or hesitant to share their true experiences due to cultural norms or fear of judgment. To address this, the study ensured confidentiality and anonymity for all respondents, creating a safe environment that encouraged honest and open dialogue during interviews and focus group discussions.

Lastly, the dynamic nature of climate change posed a challenge, as its ongoing effects could lead to continuously evolving gender roles and family dynamics that were not fully captured within the study's timeframe. To acknowledge this limitation, the study recommends that future research adopt a longitudinal approach to monitor these changes over time, thereby providing a more comprehensive understanding of how climate-induced shifts in gender roles continue to affect family cohesion.

17 Conclusion

The study reveals profound disruptions in family dynamics caused by climate-induced stressors like droughts and floods. These environmental challenges have forced a reconfiguration of traditional gender roles, with women taking on increased economic and domestic burdens. This shift has significantly strained emotional support within families, disrupted communication patterns, and heightened tensions, especially between spouses. As women assume roles as primary providers, the study found weakened spousal connections, reduced parent-child interactions, and diminished support from extended family networks. The increased prevalence of gender-based violence (GBV) and cultural resistance to changing gender norms further aggravate these issues. These findings highlight the urgent need for comprehensive strategies that address both the economic pressures and social conflicts caused by evolving gender roles. Promoting gender-sensitive policies, enhancing women's leadership, and fostering open dialogue within families are crucial steps toward strengthening family cohesion and building community resilience in the face of ongoing climate challenges.

18 Recommendations

In response to the study findings, the following recommendations are proposed to guide policy development, community initiatives, and support systems aimed at addressing the challenges posed by climate-induced gender dynamics and enhancing family cohesion:

1. **Foster gender-sensitive community engagement:** To address the strain on family cohesion caused by climate-induced role shifts, traditional leaders, local government, and NGOs should lead awareness campaigns that highlight the evolving roles of women due to droughts and floods. These campaigns should focus on promoting acceptance of women's increasing economic and decision-making responsibilities, reducing resistance to changing gender norms triggered by climate change.
2. **Strengthen communication and conflict resolution:** Climate change-related pressures, like droughts and economic hardships, have disrupted communication within families. Community organizations and mental health professionals should implement interventions that equip families with conflict resolution skills and strategies for joint decision-making. Accessible counseling services should be available to help families navigate the challenges of new gender dynamics caused by environmental stressors.
3. **Empower women economically:** In response to climate-induced economic pressures, local authorities and women's advocacy groups should develop financial literacy programs tailored to women. Empowering women to manage resources and encouraging joint financial decision-making can reduce conflicts over finances, which are exacerbated by the economic instability brought on by climate change, thereby strengthening family units.
4. **Mitigate gender-based violence (GBV):** The shift in gender roles driven by climate change has led to an increase in GBV. Policymakers, law enforcement, and healthcare providers must develop comprehensive GBV response frameworks that include shelters, legal aid, and psychological support for survivors. Community-based initiatives should address the root causes of GBV linked to economic and social pressures resulting from environmental challenges.
5. **Promote sociocultural adaptations:** Climate-induced changes in gender roles require a cultural shift toward supporting women's leadership and decision-making in response to these new realities. Traditional and religious leaders should advocate for sociocultural adaptations that validate women's evolving roles, ensuring that communities recognize the importance of women's contributions to climate resilience and family cohesion.
6. **Rebuild extended family support networks:** The breakdown of extended family support due to individual survival struggles from climate crises like droughts and floods highlights the need for collective resilience. Local governments should strengthen community-based support groups to revive traditional family networks, promoting mutual aid to help families cope with the socio-economic impacts of climate change.

7. **Integrate gender-responsive policies:** Governments and development agencies must incorporate gender-responsive policies into climate adaptation strategies. These policies should prioritize women's participation in decision-making processes and ensure that resources are allocated to support their leadership in responding to the challenges of climate change. Such initiatives will empower women to enhance household resilience and family cohesion.
8. **Provide mental health support:** The psychological toll of climate-induced role shifts requires urgent attention. NGOs and healthcare providers should deliver culturally sensitive mental health services to women facing increased economic and caregiving pressures due to climate impacts. Support systems that address emotional stress will help strengthen family cohesion and improve community resilience in the face of ongoing climate challenges.

References

1. Mondal, D. (2024). Socio-economic impacts of climate change. *International Journal on Agricultural Sciences, 15*(02), 95–96. https://doi.org/10.53390/ijas.2024.15203
2. Intergovernmental Panel on Climate Change. (2023). *Climate change 2021—The physical science basis*. https://doi.org/10.1017/9781009157896
3. Cerrone, E., & Rodriguez Ponce de León, C. (2025). Género y adversidad ambiental: Desigualdades agudizadas. *Seminario Internacional de Investigación En Urbanismo*. https://doi.org/10.5821/siiu.12883
4. Guha, S. (2024). Women, gender equality, and environmental justice. *Advances in Religious and Cultural Studies (ARCS) Book Series*, 245–268. https://doi.org/10.4018/979-8-3693-6069-9.ch013
5. Climate Change 2022—Impacts, Adaptation and Vulnerability. (2023). https://doi.org/10.1017/9781009325844
6. Mukisa, A., Muhanguzi, F. K., & Boonabaana, B. (2023). The effects of climate change on gender roles among agro-pastoral farmers in Nabilatuk district, Karamoja subregion, North Eastern Uganda. *Frontiers in Human Dynamics*. https://doi.org/10.3389/fhumd.2023.1092241
7. Satapathy, S. S. (2024). Climate change and feminization of agriculture in India. In *Advances in Religious and Cultural Studies (ARCS) Book Series* (pp. 405–416). https://doi.org/10.4018/979-8-3693-6069-9.ch020
8. Karmakar, S. (2022). Climate change and everyday-life: Negotiation of women in the sundarbans, india. *Man, Environment and Society, 3*(1), 49–64. https://doi.org/10.47509/mes.2022.v03i01.04
9. Nagasha, J. I., Mugisha, L., Kaase-Bwanga, E., Onyuth, H., & Ocaido, M. (2019). Effect of climate change on gender roles among communities surrounding Lake Mburo National Park, Uganda. *Emerald Open Research, 1*, 7. https://doi.org/10.12688/EMERALDOPENRES.12953.2
10. Niemann, J., El-Mahdi, M., Samuelsen, H., & Tersbøl, B. P. (2024). Gender relations and decision-making on climate change adaptation in rural East African households: A qualitative systematic review. *PLOS Climate*. https://doi.org/10.1371/journal.pclm.0000279
11. Nagendiran, K., & Sathya, R. (2024). Economic impacts of climate change on Indian farmers and rural livelihoods. *International Journal For Multidisciplinary Research, 6*(6). https://doi.org/10.36948/ijfmr.2024.v06i06.33839

12. Southard, E. M. L., & Randell, H. E. (2022). Climate change, agrarian distress, and the feminization of agriculture in South Asia. *Rural Sociology, 87*(3), 873–900. https://doi.org/10.1111/ruso.12439
13. Deshar, R., & Koirala, M. (2020). *Climate change and gender policy* (pp. 411–422). Springer. https://doi.org/10.1007/978-981-13-9570-3_14
14. Pearse, R. (2017). Gender and climate change. *Wiley Interdisciplinary Reviews: Climate Change, 8*(2). https://doi.org/10.1002/WCC.451
15. Acosta, M., van Wessel, M., van Bommel, S., & Feindt, P. H. (2021). Examining the promise of 'the local' for improving gender equality in agriculture and climate change adaptation. *Third World Quarterly, 42*(6), 1135–1156. https://doi.org/10.1080/01436597.2021.1882845
16. Andersen, L. E., Verner, D., & Wiebelt, M. (2014). Gender and climate change in Latin America: An analysis of vulnerability, adaptation and resilience based on household surveys. *Research Papers in Economics*. https://ideas.repec.org/p/adv/wpaper/201408.html
17. Aipira, C., Kidd, A., & Morioka, K. (2017). *Climate change adaptation in Pacific countries: Fostering resilience through gender equality* (pp. 225–239). Springer. https://doi.org/10.1007/978-3-319-50094-2_13
18. Alston, M., & Fuller, S. (2023). Women and climate change in Vanuatu, Pacific Islands Region. *Gender Place and Culture*, 1–22. https://doi.org/10.1080/0966369x.2023.2229530
19. Gicheru, M., Mwenda, M. J., & Omwami, D. O. (2024). Gender and climate change: The role of women in climate change processes. *Asian Journal of Geographical Research*. https://doi.org/10.9734/ajgr/2024/v7i1210
20. Saif, F., & Raza, A. (2022). Women and climate change. *Pakistan Journal of Social Research, 04*(03), 737–743. https://doi.org/10.52567/pjsr.v4i03.763
21. Nombo, C. I., Kabote, S. J., Mamiro, D. P., Synnevåg, G., Mattee, A. Z., Urassa, J. K., & Matata, L. M. (2015). *Adaptation to climate change: Changing gender relations in the Meatu and Iramba Districts in Tanzania* (pp. 587–599). Springer. https://doi.org/10.1007/978-3-319-09360-4_31
22. Awiti, A. (2022). Climate change and gender in Africa: A review of impact and gender-responsive solutions. *Frontiers in Climate, 4*. https://doi.org/10.3389/fclim.2022.895950
23. Chikuruwo, R. E. (2024). Navigating shifting currents: Gendered vulnerabilities and climate change in the Lake Chad Basin. *Agenda*. https://doi.org/10.1080/10130950.2023.2330801
24. Defina, D., Hastuti, D., Rizkillah, R., Muflikhati, I., & Sheriffdeen, M. (2024). Rippling tides: Understanding climate change impacts, gender roles, communication and family dynamics in indonesian coastal communities. *Migration Letters*. https://doi.org/10.59670/ml.v21i5.9611
25. Nunow, A. A. (2024). The nexus between climate change and livelihoods in arid and semi-arid (ASAL) areas of Kenya: Evidence from the literature. *International Journal of Social Science and Human Research, 7*(06). https://doi.org/10.47191/ijsshr/v7-i06-113
26. Glazebrook, T. (2017). *Women and climate change* (pp. 545–557). Routledge. https://doi.org/10.4324/9781315097176-76
27. Khan, M. A. (2024). Reviewing the impacts of climate change on women. *International Journal of Advanced Scientific Research and Management, 9*(8), 1–11. https://doi.org/10.36282/ijasrm/9.8.2024.1946
28. Pettengell, C. (2015). *Africa's smallholders adapting to climate change: The need for national governments and international climate finance to support women producers*. https://oxfamilibrary.openrepository.com/handle/10546/579620
29. Bomuhangi, A., Nabanoga, G., Namaalwa, J. J., Jacobson, M. G., & Gombya-Ssembajjwe, W. (2016). Gendered decision making and adaptation to climate change in Mt. Elgon Region, Eastern Uganda. *International Research Journal of Environmental Sciences and Studies, 1*(1). https://www.cifor.org/publications/pdf_files/articles/ABomuhangi1601.pdf
30. Wanjara, A. O., & Ogembo, P. O. (2023). Impact of climate change on health and livelihoods of pastoral communities in Kenya: A case of north eastern region. *Indonesian Journal of Social and Environmental Issues*. https://doi.org/10.47540/ijsei.v4i3.1078
31. Fakir, A. M. S., Fakir, A. M. S., & Abedin, N. (2021). Empowered by absence: Does male out-migration empower female household heads left behind? *Journal of International Migration and Integration, 22*(2), 503–527. https://doi.org/10.1007/S12134-019-00754-0

32. Ngigi, M. W., Mueller, U., & Birner, R. (2016). *ZEF-Discussion Papers on Development Policy No. 210*.
33. Onyango, E. A., Banwell, N., Mose, T. M., & Ogallo, L. A. (2023). Editorial: Gendered impacts of climate change: Women and transformative research, policy and practice. *Frontiers in Climate, 5*. https://doi.org/10.3389/fclim.2023.1215661
34. Ume, C. O., Opata, P. I., & Onyekuru, A. N. (2021). *Gender and climate change adaptation among rural households in Nigeria* (pp. 2099–2115). Springer. https://doi.org/10.1007/978-3-030-45106-6_182
35. Moser, C. O. N. (1989). Gender planning in the third world: Meeting practical and strategic gender needs. *World Development, 17*(11), 1799–1825.
36. Allison, K. H. (2023). *Gender, and development*. https://doi.org/10.31219/osf.io/utkv4
37. Robinson, Z. F. (2018). *Intersectionality and gender theory* (pp. 69–80). Springer. https://doi.org/10.1007/978-3-319-76333-0_5
38. Dhuli, B., Dhamo, A., & Dhamo, I. (2023). Gender equality as a necessary approach for the country's development process and for gender integration. *Migration Letters*. https://doi.org/10.59670/ml.v21i2.6433
39. Ghosh, A., & Ghosh, A. (2024). *Gender inequality: Examining the social, economic, and cultural factors that contribute to gender disparities and exploring strategies for achieving gender equality* (pp. 128–143). https://doi.org/10.58532/v3baso14p3ch2
40. Costache, A. (2023). *Intersectionality and climate change: A review of the literature between 2014 and 2023*. https://doi.org/10.24940/theijhss/2023/v11/i9/hs2309-017
41. Nosheen, M., Iqbal, J., & Ahmad, S. (2023). Economic empowerment of women through climate change mitigation. *Journal of Cleaner Production*. https://doi.org/10.1016/j.jclepro.2023.138480
42. Adamaagashi, I. P., Onoh, U. C., Ogunade, J. A., Ayanlowo, E. J., & Nwimo, C. P. (2023). Examining the social and economic impact of climate change on rural communities and the role of global governance in addressing these challenges. *IIARD International Journal of Geography and Environmental Management*. https://doi.org/10.56201/ijgem.v9.no4.2023.pg1.18
43. Boetto, H., & McKinnon, J. (2013). Rural women and climate change: A gender-inclusive perspective. *Australian Social Work, 66*(2), 234–247. https://doi.org/10.1080/0312407X.2013.780630
44. Bista, R. B. (2023). Assessment of climate change induced household vulnerability in Nepal: Application of index method. *Arrus Journal of Social Sciences and Humanities, 3*(1), 9–20. https://doi.org/10.35877/soshum1151
45. Walker, T., Kawasoe, Y., & Shrestha, J. (2019). *Risk and vulnerability in Nepal: Findings from the household risk and vulnerability survey*. 1–92. https://documents.worldbank.org/curated/en/318281581716869564/Risk-and-Vulnerability-in-Nepal-Findings-from-the-Household-Risk-and-Vulnerability-Survey
46. Smith-Greenaway, E., & Trinitapoli, J. (2014). Polygynous contexts, family structure, and infant mortality in sub-Saharan Africa. *Demography, 51*(2), 341–366. https://doi.org/10.1007/S13524-013-0262-9
47. Novillo Rameix, N. N. (2018). Cambio climático y conflictos socioambientales en ciudades intermedias de América Latina y el Caribe/Climate change and environmental social conflicts in intermediary cities of Latin America and the Caribbean. *Letras Verdes: Revista Latinoamericana de Estudios Socioambientales, 24*, 124–142. https://doi.org/10.17141/LETRASVERDES.24.2018.3323
48. Leichenko, R., & Silva, J. A. (2024). *Climate change and poverty* (pp. 374–387). https://doi.org/10.4324/9781003291206-31
49. Zulfikri, H. F. B. (2024). Empowering through education: Climate change and financial literacy. *Social Science Research Network*. https://doi.org/10.2139/ssrn.4844640
50. Chaity, Z. Y. (2022). Climate-induced migrations and changing 'households' in Bangladesh: An analysis of new 'householding' structures from gender perspectives. *Global Journal of Human Social Science*, 31–36. https://doi.org/10.34257/gjhsshvol22is4pg31

51. Onwunta, O. E. K. (2022). *Men and climate change: Some thoughts on South Africa and Bangladesh*. 1–17. https://doi.org/10.1080/18902138.2022.2077082
52. Leal Filho, W., Ayal, D. Y., Chamma, D. D., Kovaleva, M., Alverio, G. N., Nzengya, D. M., Mucova, S., Kalungu, J. W., & Nagy, G. J. (2023). *Assessing causes and implications of climate-induced migration in Kenya and Ethiopia*. https://doi.org/10.1016/j.envsci.2023.103577
53. Behere, A. P., & Dell, M. L. (2023). *At the intersection of climate change and children's mental health: A global perspective through visual narratives*. https://doi.org/10.1016/j.jaac.2023.07.334
54. Herman, M. (2016). *Women and climatic changes in Ethiopia: A gendered assessment* (Vol. 5). Centre français des études éthiopiennes. https://doi.org/10.4000/BOOKS.CFEE.417
55. Mahapatra, B., Chaudhuri, T., & Saggurti, N. (2022). Climate change vulnerability, and health of women and children: Evidence from India using district level data. *International Journal of Gynecology & Obstetrics, 160*(2), 437–446. https://doi.org/10.1002/ijgo.14515
56. Randell, H., & Gray, C. (2019). Climate change and educational attainment in the global tropics. *Proceedings of the National Academy of Sciences of the United States of America, 116*(18), 8840–8845. https://doi.org/10.1073/PNAS.1817480116
57. Sanson, A., Padilla Malca, K., Van Hoorn, J., & Burke, S. E. L. (2023). Children and climate change. *The Brown University Child and Adolescent Behavior Letter, 39*(4), 8. https://doi.org/10.1002/cbl.30703
58. Mannell, J., Brown, L. J., Jordaan, E., Hatcher, A. M., & Gibbs, A. (2024). *The impact of climate shocks due to climate change on intimate partner violence: A structural equation model of data from 156 countries*. https://doi.org/10.2139/ssrn.4800400
59. Segnon, A. C., Totin, E., Zougmoré, R. B., Lokossou, J., Thompson-Hall, M., Ofori, B. O., Achigan-Dako, E. G., & Gordon, C. P. (2021). Differential household vulnerability to climatic and non-climatic stressors in semi-arid areas of Mali, West Africa. *Climate and Development, 13*(8), 697–712. https://doi.org/10.1080/17565529.2020.1855097
60. Nishat, K. J., & Rahman, M. S. (2019). *Disaster, vulnerability, and violence against women: Global findings and a research agenda for Bangladesh* (Vol. 4, pp. 1859–1874). IGI Global. https://doi.org/10.4018/978-1-5225-6912-1.CH098
61. Schumacher, J. A., Coffey, S. F., Norris, F. H., Tracy, M., Clements, K., & Galea, S. (2010). Intimate partner violence and Hurricane Katrina: Predictors and associated mental health outcomes. *Violence & Victims, 25*(5), 588–603. https://doi.org/10.1891/0886-6708.25.5.588
62. Sety, M., James, K., & Breckenridge, J. (2014). *Understanding the risk of domestic violence during and post natural disasters: Literature review* (pp. 99–111). Springer. https://doi.org/10.1007/978-3-319-05882-5_5
63. Dewan, C. (2024). Affective assemblages of kinship and single mothers' labour migration from a 'climate hotspot'. *Journal of the Royal Anthropological Institute*. https://doi.org/10.1111/1467-9655.14238
64. Karpe, P. (2022). *Climate change, conflict, and prosocial behavior in southwestern Bangladesh: Implications for environmental justice* (pp. 349–369). https://doi.org/10.1007/978-981-19-1987-9_17
65. Osei-Amponsah, C., Quarmine, W., & Okem, A. (2023). Understanding climate-induced migration in West Africa through the social transformation lens. *Frontiers in Sociology*. https://doi.org/10.3389/fsoc.2023.1173395
66. Reyes-Díaz, M. (2022). *Climate-induced migration in West Africa* (pp. 79–105). Springer eBooks. https://doi.org/10.1007/978-3-030-97322-3_5
67. Dzah, E. D. E. (2011). *Gender dynamics of climate change in Ghana: An intersectional perspective*. https://thesis.eur.nl/pub/10591/Ellen_Dzah_WGD_2011.pdf
68. Goudet, J.-M., Owais, H. M. A., Ahmed, H. U., & Ridde, V. (2023). Intersecting vulnerabilities: Unraveling the complex relationship between women, climate change, and mental health in Bangladeshi communities. *medRxiv*. https://doi.org/10.1101/2023.06.06.23290721
69. Yimam, D., & Holvoet, N. (2023). Unpacking the invisible complex realities: Intersections of gender and marital status in determining the intrinsic vulnerability of smallholder farmers to

climate change in Northwestern Ethiopia. *Climate and Development*. https://doi.org/10.1080/17565529.2023.2246038

70. Collins, J. (2014). A rising tide in Bangladesh: Livelihood adaptation to climate stress. *Australian Geographer, 45*(3), 289–307. https://doi.org/10.1080/00049182.2014.929999
71. Iliyasu, J., & Mamman, S. O. (2024). Climate change and human economic misery in West Africa: Evidence from a tail-dependence model. *Climate and Development*, 1–9. https://doi.org/10.1080/17565529.2024.2437123
72. Burke, S. E. L., Sanson, A., & Van Hoorn, J. (2018). The psychological effects of climate change on Children. *Current Psychiatry Reports, 20*(5), 35. https://doi.org/10.1007/S11920-018-0896-9
73. Ahmed, A., Ahmed, A., Lawson, E. T., Mensah, A., Gordon, C. P., & Padgham, J. (2016). Adaptation to climate change or non-climatic stressors in semi-arid regions? Evidence of gender differentiation in three agrarian districts of Ghana. *Environmental Development, 20*, 45–58. https://doi.org/10.1016/J.ENVDEV.2016.08.002
74. Djoudi, H., Djoudi, H., Brockhaus, M., & Brockhaus, M. (2011). Is adaptation to climate change gender neutral? Lessons from communities dependent on livestock and forests in northern Mali. *International Forestry Review, 13*(2), 123–135. https://doi.org/10.1505/146554811797406606
75. Esse, K. K. J. B., & Yao, J. (2021). *Changement climatique et réfugiés en Afrique de l'Ouest (1990–2020)*. https://doi.org/10.61585/pud-asasx-v1n311
76. Momtaz, S., & Asaduzzaman, M. (2018). *Climate change impacts and women's livelihood: Vulnerability in developing countries*. https://www.taylorfrancis.com/books/mono/10.4324/9780429462474/climate-change-impacts-women-livelihood-salim-momtaz-muhammad-asaduzzaman
77. Sujakhu, N. M., Ranjitkar, S., He, J., Schmidt-Vogt, D., Su, Y., & Xu, J. (2019). Assessing the livelihood vulnerability of rural indigenous households to climate changes in central Nepal, Himalaya. *Sustainability, 11*(10), 2977. https://doi.org/10.3390/SU11102977
78. Margaglio, G., & Chesnier, A. (2019). *Linking climate change, migration patterns and vulnerability: The case of Ndem, Senegal*. https://lup.lub.lu.se/student-papers/record/8995448/file/8995979.pdf
79. Epstein, A., Bendavid, E., Nash, D., Charlebois, E. D., & Weiser, S. D. (2020). Drought and intimate partner violence towards women in 19 countries in sub-Saharan Africa during 2011-2018: A population-based study. *PLOS Medicine, 17*(3). https://doi.org/10.1371/JOURNAL.PMED.1003064
80. Kangoma, M., Lubanga, F. A., & Bwanali, A. N. (2023). *Mental health impact of the 2023 Tropical Cyclone Freddy and Southern region mudslides on Malawian women: The neglected humanitarian crisis*. https://doi.org/10.1097/gh9.0000000000000306
81. Musa, F. B., Katundu, M., Lewis, L. A., & Munthali, A. (2024). Gender and livelihood assets: Assessing climate change resilience in phalombe district—Malawi. *Environmental and Sustainability Indicators*. https://doi.org/10.1016/j.indic.2024.100347
82. Mendoza, M. E. T., Naret, H., Ballaran, V. G., & Arias, J. K. B. (2014). Assessing vulnerability to climate change impacts in Cambodia, the Philippines and Vietnam: An analysis at the commune and household level. *Journal of Environmental Science and Management, 17*(2). https://ovcre.uplb.edu.ph/journals-uplb/index.php/JESAM/article/view/190
83. Sorensen, C., Sorensen, C., Saunik, S., Sehgal, M., Tewary, A., Govindan, M., Lemery, J., & Balbus, J. (2018). Climate change and women's health: Impacts and opportunities in India. *Geohealth, 2*(10), 283–297. https://doi.org/10.1029/2018GH000163
84. Balehey, S., Tesfay, G., Balehegn, M., & Balehegn, M. (2018). Traditional gender inequalities limit pastoral women's opportunities for adaptation to climate change: Evidence from the Afar pastoralists of Ethiopia. *Pastoralism, 8*(1), 1–14. https://doi.org/10.1186/S13570-018-0129-1
85. Huq, H., Rahman, M. M., & Hossen, M. A. (2023). Women-led climate change adaptation: A study in northwestern Bangladesh. *Sustainability*. https://doi.org/10.3390/su152015041
86. Djoudi, H. (2015). *At the intersection of inequities: Lessons learned from CIFOR's work on gender and climate change adaptation in West Africa*. https://www.cifor.org/publications/pdf_files/Brief/5899-GenderClimateBrief.pdf

87. Memon, F. S. (2020). Climate change and violence against women: Study of a flood-affected population in the rural area of Sindh, Pakistan. *Pakistan Journal of Women's Studies Alam-e-Niswan, 27*(1), 65–85. https://doi.org/10.46521/PJWS.027.01.0039
88. Nsengiyumva, G., Clarkson, G. J. J., Gumucio, T., Dorward, P., & Ingabire, C. M. (2022). Exploring the effects of a participatory climate services approach on smallholder decision-making in Rwanda using a gender lens. *Gender, Technology and Development, 26*(3), 544–563. https://doi.org/10.1080/09718524.2022.2144102
89. Aryal, J. P., Sapkota, T. B., Rahut, D. B., Gartaula, H., & Stirling, C. M. (2022). Gender and climate change adaptation: A case of Ethiopian farmers. *Natural Resources Forum, 46*(3), 263–288. https://doi.org/10.1111/1477-8947.12259
90. Jørstad, H., & Webersik, C. (2016). Vulnerability to climate change and adaptation strategies of local communities in Malawi: Experiences of women fish-processing groups in the Lake Chilwa Basin. *Earth System Dynamics Discussions, 7*(4), 977–989. https://doi.org/10.5194/ESD-7-977-2016
91. Aman, S., Tameez-ud-din, A., Afraz, N., Ehtisham, M., & Khan, J. (2024). Climate change mitigation and women's economic empowerment. *Journal of Health and Climate Change, 3*(1). https://doi.org/10.37939/jhcc.v3i1.10
92. Andersen, L. E., Verner, D., & Wiebelt, M. (2017). Gender and climate change in Latin America: An analysis of vulnerability, adaptation and resilience based on household surveys. *Journal of International Development, 29*(7), 857–876. https://doi.org/10.1002/JID.3259

Restoration of Ecosystems and Mitigation Strategies

The Role of Vegetation Health and Nature-Based Solutions in Mitigating Climate Change in the River Isiukhu Basin

Samuel Abuyeka Tela and Nelly Nambande Masayi

Abstract

Assessment of how well vegetation health and nature-based solutions can offset climate change is vital for mitigation and adaptation planning but has rarely been done. This study assessed the role of vegetation health and NbS in mitigating climate change effects using the Normalised Difference Vegetation Index and Normalised Different Bare Surface Index, which were derived from Google Earth Engine and ArcGIS Pro 3.2. Precipitation and temperature data were collected from Climate Hazards Group InfraRed Precipitation with Station data (CHIRPS) and TERRACLIMATE, respectively. Pearson correlation explored the relationship between remote sensing indices, NbS and climate data. Four afforestation NbS case study projects were noted. NDVI increased by 35% over the study period. NDVI increase was noted (1990–2013), while decline was noted (2013–2023). NDBSI decreased (1990–2013) and increased (2013–2023). Precipitation increased by 22.7% (427.3 mm) (1990–2023), while temperature increased by 7.43% (1.5 °C) over the same period. Positive relationship between precipitation and NDVI ($r = 0.5549$) was noted, while a negative correlation between precipitation and NDBSI ($r = -0.139$) was realised. Temperature was both positively correlated with NDVI ($r = 0.8237$) and NDBSI ($r = 0.1916$).

S. A. Tela (✉)
Department of Social Studies, School of Education, Humanities and Social Sciences, Jaramogi Oginga Odinga University of Science and Technology, Bondo, Kenya

Department of Geography, School of Law and Social Sciences, Kenyatta University, Nairobi, Kenya

N. N. Masayi
Faculty of Arts and Social Sciences, Kibabii University, Bungoma, Kenya

Institute for Climate Change Adaptation, University of Nairobi, Nairobi, Kenya

N. Ronoh, C. Rono (eds.), *Rethinking Our Ecosystems*, Springer Proceedings in Geography and Sustainability, https://doi.org/10.1007/978-3-031-95781-9_3

There was a strong positive significant correlation between NDVI and precipitation before and after the introduction of various NbS ($r = 0.9692$, $p > 0.05$). Therefore, vegetation health and NbS controlled the climatic conditions of the basin. This study prioritises the adoption of NbS for climate change mitigation. The study findings can be used as a reference for measuring the effectiveness of NbS in mitigating climate change in the world.

Keywords
Nature-based solution · Normalized difference bare soil index · Climate change · Normalized difference vegetation index

1 Introduction

River basins are important hotspots of biodiversity and social-economic revolution. However, they have become hotspots of degradation by human and climatic pressures [6, 19]. Vegetation cover changes because natural and anthropogenic pressures accelerate the ecosystem stress, and this could have devastating effects on communities that rely on them for livelihoods [5]. Unsustainable land use in sub-Saharan Africa arising from demographic changes is the major precursor catchment degradation [25]. This degradation has devastating effects on the local communities' social-economic activities and environment, consequently contributing to climate change.

River catchment degradation causes substantial increase in bare surfaces as well as decline of green vegetation cover [20]. As a result, it disrupts energy and water balance by modifying the earth surface LST and precipitation patterns. Continual land cover management and conservation has the potential to reduce the negative consequences of climate change and ensures the human ecological health and economic stability of communities [20].

According to the UNEA-5 resolution, NbS are 'actions to protect, conserve, restore, sustainably use and manage natural or modified terrestrial, water ecosystems, by addressing social, ecological, economic and environmental challenges effectively and adaptively, while providing human well-being' [12]. Globally, NbS have been implemented in various ecosystems to address specific societal challenge and enhance several other benefits. NbS can therefore represent a new generation of approaches that are more flexible and adaptable to the global environmental changes [22]. They could therefore be the most effective solution to the river basin ecosystem conservation, given the degradation that they face from climate change and anthropogenic activities.

Among the major NbS utilised across the globe include planting of trees to reduce the effects of floods, enhance groundwater recharge and reduce global warming [11]. With strong global warming in the past, present and in the projected future, increased vegetation cover in Kenya could be considered as an NbS that could therefore increase precipitation, reduce temperature and modify vegetation health [17]. Afforestation could therefore be an effective NbS that could consequently moderate the effects of climate change and improve vegetation health in Kenya.

Land cover conversion rates and their consequences are easily monitored and detected by remote sensing indices such as NDVI, NDBSI, Normalised Difference Built-up index (NDBI) and Normalized Difference Water Index (NDWI) [4, 5, 14, 23]. NDVI, which is used to determine the greenness of vegetation, acts as the indicator of vegetation growth, health and ecosystem change [5]. NDBSI acts as an indicator of levels of land degradation due to anthropogenic and climatic drivers [10]. NDBI is valuable for monitoring and analysing urban land cover changes and the spatial distribution of built-up areas, while NDWI monitors water content availability and its spatial changes. Therefore, remote sensing indices and GISs are vital in water catchment mapping [3]. These enable decision-makers to adopt equitable and sustainable land management and conservation practices aimed at improving vegetation cover and reducing bare surface cover and enhancing community livelihood [10].

Temperature which varies in relation to human land conversion processes is an effective component for tracking an area's heat balance that influences the earth's climate change and other key physical and chemical processes [7, 16]. Precipitation and temperature are vital climatic variables in environmental research and respond differently to temporal changes in vegetation composition, bare surfaces, built-up environments and other anthropogenic land conversion processes [3, 9, 14]. NDVI has a stronger correlation with temperature and precipitation than NDBSI [10]. The nature and strength of the relationship between NDVI/NDBSI and temperature/precipitation heavily depend on the intensity of human activities, space and time. In most cases, a negative relationship generally exists between NDVI and temperature in tropical environments [27].

River Isiukhu is the major river in River Isiukhu basin and is a tributary of river Nzoia which drains into Lake Victoria, the second largest freshwater lake in the world. Despite this river basin being a significant biodiversity hotspot and a source of livelihood, it has also become a hotspot of degradation [24]. Various studies have been documented on River Isiukhu, but little has been documented on the role of NbS and vegetation health in mitigating climate change in River Isiukhu basin. The results of the study will help in the planning of NbS intervention for climate change mitigation in the basin. The study objectives were (1) to analyse the extent of decadal changes in NDVI, NDBSI, temperature and precipitation patterns in River Isiukhu basin for 1990, 2003, 2013 and 2023; (2) to analyse the extent of changes in NDVI in relation to different introduction periods of NbS projects and (3) to determine the relationship between NDVI/NDBSI and temperature/precipitation in relation.

2 Materials and Methods

2.1 The Study Area

Figure 1 shows the study area (River Isiukhu Basin), which is located in western Kenya. The latitudinal extent of the area is 0° 15′-0° 25′ north, while longitudinal extent of the basin is 34° 40′-34° 50′ east of the equator. The river basin covers approximately 683 km^2. The upper and middle section of the river basin is covered by dense forest (Kakamega forest and North Nandi forest), while the lower section is urbanised and dominated by agricultural surfaces. River Isiukhu is the main river in the basin. Other rivers in the basin include Lugusida, Shirerwa, Koa, Nandamungo and Kabkalet. The river basin experiences a typical equatorial climate associated with high rainfall and high temperatures. The basin elevation ranges between 1240 and 2000 meters above sea level. General land cover within the river basin varies in relation to natural forest, grassland, agriculture and urban settlement.

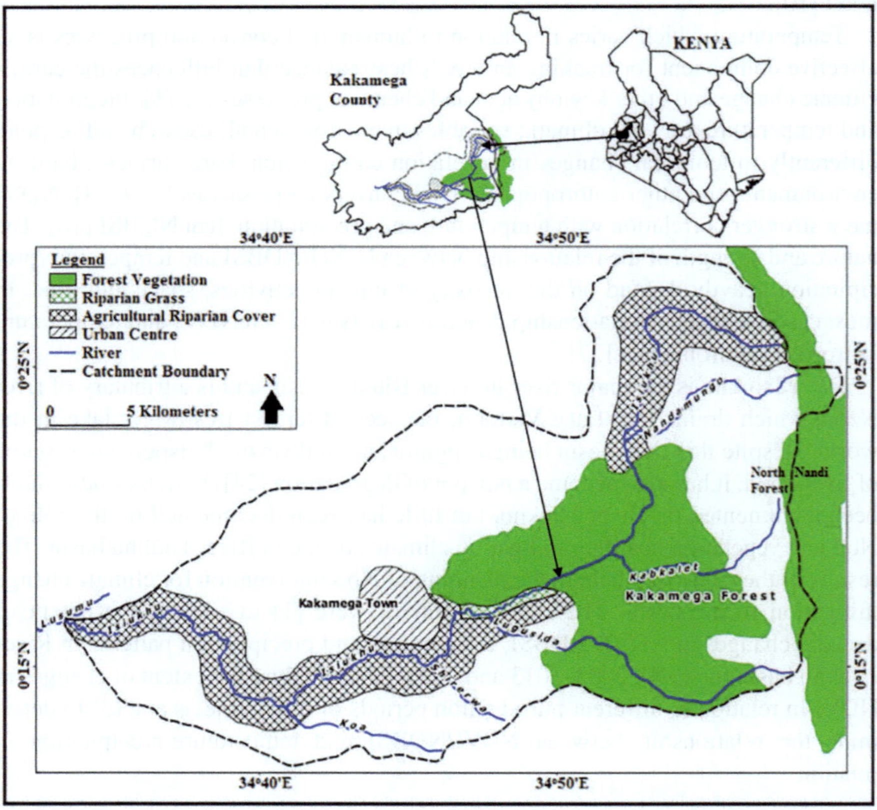

Fig. 1 River Isiukhu basin. Source: Topographic Sheet for Eldoret (NA-36-16) and Google Earth Satellite Map (2020)

2.2 Data Collection and Analysis

2.2.1 NDVI and NDBSI

To establish variations in vegetation health, satellite images from the United States Geological Survey (USGS) website were collected and analysed for the year 1990, 2003, 2013 and 2023. Landsat 4 images were used, and a median of one image was derived for the year 1990 and 2003 using google Earth Engine. For 2013 and 2023, Landsat 8 images were used. The images underwent pre-processing which included cloud removal, clipping and compositing based on the area of study using Google Earth Engine and ArcGIS Pro.

NDVI was used to determine the vegetation health. NDVI uses light reflectance in near-infrared (NIR) and visible wavelengths to determine vegetation viability and health. NDVI for 1990 and 2003 used Landsat 4, while NDVI for 2013 and 2023 used Landsat 8. NDVI was derived using the following formula:

$$NDVI = (NIR\text{-RED})/(NIR\text{+RED})$$

For 1990 and 2003 imageries:

NIR = band 4.
RED = band 3.

As for the imageries for 2013 and 2023:

NIR = band 5.
RED = band 4.

It involved subtracting visible red reflectance from NIR reflectance and dividing it with the sum of both red reflectance and NIR reflectance. NDVI is a popular satellite-based index used for monitoring vegetation health over large areas [13]. Comparative studies of prolonged-time series of NDVI can be used to give helpful evidence for tracking regions with and without NbS intervention. The NDVI was delivered as a single-band product with values ranging from −1 to +1, with bare soil having a value around zero and plants having a positive value [15].

NDBSI was used to establish the extent of vegetation loss and land degradation in the region. NDBSI uses visible red and blue wavelengths, light reflectance in near-infrared (NIR) and short-wave infrared band (SWIR) to determine land degradation and vegetation health. NDBSI for 1990 and 2003 used Landsat 4 imageries, while NDBSI for 2013 and 2023 used Landsat 8. NDBSI was derived using the following formula:

$$NDBSI = (RED+SWIR) - (NIR+BLUE)/(RED+SWIR)+(NIR+BLUE)$$

For 1990 and 2003 imageries:

RED = band 3.
SWIR = band 5.

NIR = band 4.
BLUE = band 1.

For 2013 and 2023 imageries:

RED = band 4.
SWIR = band 6.
NIR = band 5.
BLUE = band 2.

It involved adding visible red reflectance and short-wave infrared band (SWIR) and then subtracting it from the sum of light reflectance in near-infrared (NIR) and visible blue reflectance. The result is then divided by the sum of visible red reflectance and short-wave infrared band (SWIR) and light reflectance in near-infrared (NIR) and visible blue reflectance. The NDBSI values ranged from −1 to 1 with −1 signifying areas that have dense vegetation cover, while 1 indicating areas that are heavily degraded and with barely no vegetation cover.

2.2.2 Nature-Based Solution Data

Data on available nature based solutions within the river basin was established through field surveys, data collection, key informant interviews and literature review.

2.2.3 Climatic Data Collection

Data on precipitation was collected from Climate Hazards Group InfraRed Precipitation with Station (CHIRPS) data. Temperature data was collected from TERRACLIMATE data (https://developers.google.com/earthengine/datasets/catalog/IDAHO_EPSCOR_TERRACLIMATE). Climate data was collected for the year 1990, 2003, 2013 and 2023. It was downloaded and analysed into mean annual precipitation, mean daily precipitation, mean monthly temperature and mean annual temperature for the stated period of time.

3 Results and Discussions

3.1 NDVI and NDBSI Changes and Trends in the River Basin

NDVI maps for 1990, 2003, 2013 and 2023 were prepared to show the changes in the river basin with clarity. Results show that there was a decadal increase in vegetation health index (Fig. 2). NDVI maps for 1990, 2003, 2013 and 2023 revealed mean NDVI of 0.44, 0.6, 0.61 and 0.597, respectively. NDVI in Isiukhu basin increased by 35% between 1990 and 2023, 36.36% between 1990 and 2003 and 1.67% between 2003 and 2013. However, an alarming signal was the decrease in NDVI by 2.13% between 2013 and 2023. The decreased NDVI could be attributed to the loss of green vegetation cover due to factors, such as deforestation, adverse climatic conditions such as global warming and a decline in crop productivity. This

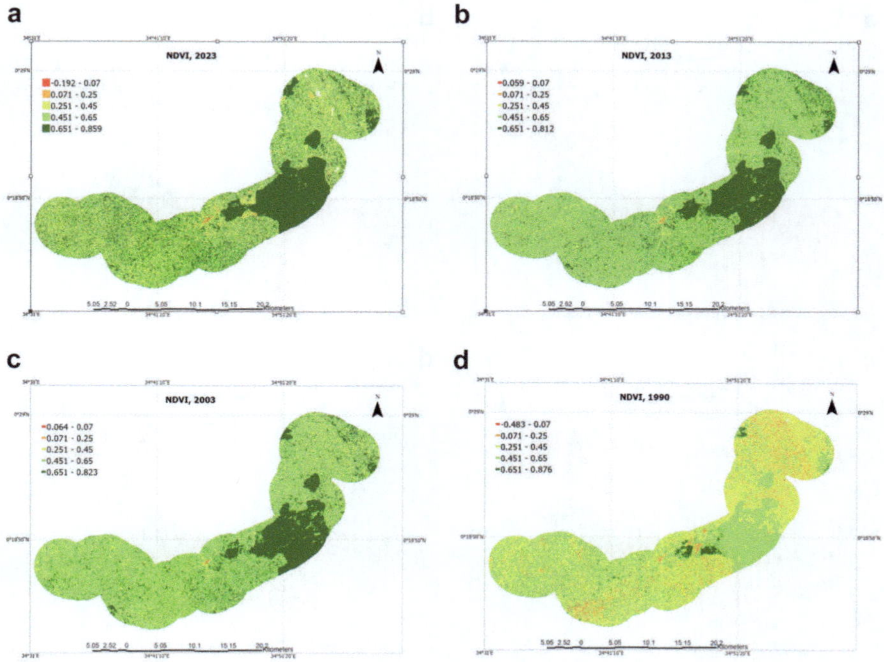

Fig. 2 (**a**) NDVI status (2023) (**b**) NDVI status (2013) (**c**) NDVI status (2003) (**d**) NDVI status (1990)

change might also be attributed to the observed increased temperature of 3.83% (0.8 °C) between 2013 and 2023. The general increase in NDVI over 33 years can either signify increased land cover management and conservation efforts such as afforestation, re-afforestation and forest protection carried out by various NbS organisations. However, increase in NDVI can also signify increased crop farming. This should be considered negative since agricultural green fields cannot replace the role that the natural vegetation has on climate change mitigation as well as maintaining the river basin ecosystem functions.

NDBSI in the river basin decreased progressively between 1990 and 2013 and increased between 2013 and 2023 as shown in the figure below (Fig. 3). NDBSI maps for 1990, 2003, 2013 and 2023 revealed mean NDBSI of −0.09, −0.1, −0.18 and − 0.08, respectively. NDBSI in the basin increased by −11.11% between 1990 and 2023. NDBSI decreased by 11.1% between 1990 and 2003, decreased by 80% between 2003 and 2013 and increased by −55.56% between 2013 and 2023. The decrease in NDBSI between 1990 and 2013 could be attributed to increased vegetation cover in the region as propagated by the various NbS programs in the region. However, an increase in NDBSI between 2013 and 2023 could signify an anthropogenic or natural change responsible for increase in the bareness of surfaces due to low vegetation cover. This can be a result of factors such as deforestation, urbanisation, expansion of agronomic activities and increased land degradation. NDBSI is

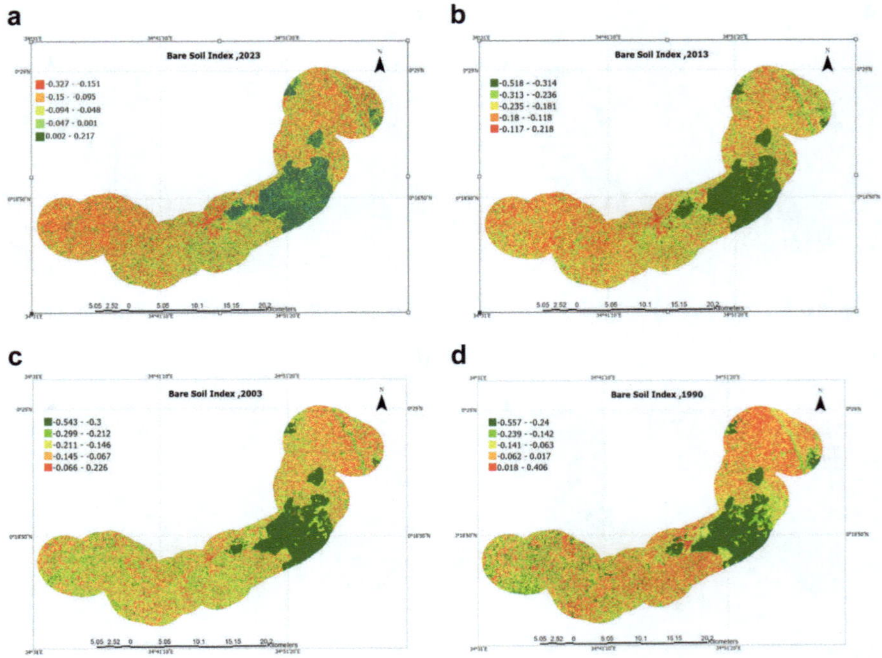

Fig. 3 (**a**) NDBSI status (2023) (**b**) NDBSI status (2013) (**c**) NDBSI status (2003) (**d**) NDBSI status (1990)

vital in understanding land cover dynamics which helps to support sustainable land use and conservation measures [8, 10].

3.2 Nature-Based Solutions in the Study Area

The major NbS recorded in the region was the Plantation Establishment and Livelihood Improvement Scheme (PELIS). It was established after enactment of the Forest Act (2005) [1]. PELIS is an agroforestry system that has been adopted by the Kenyan government in effort to protect and conserve forest ecosystems such as Kakamega forest. Through this system, the local community are allowed to grow crops under agroforestry in degraded regions within the forest. The farmers are allowed to tend their crops even as they watch over the tree seedling. Once the crops are harvested, it is expected that the trees are left to grow as a rehabilitation strategy within the deforested region.

The second NbS in the region is Muliru Farmers Conservation Group that was founded in 1997 [26]. This is a community-based organisation located near Kakamega Forest in western Kenya. The group generates income through the commercial cultivation and secondary processing of an indigenous medicinal plant, to produce the medicinal products. The enterprise reduces pressure on the Kakamega

Forest by offering an alternative to the exploitation of forest resources, while the commercialisation of the medicinal plant has heightened local appreciation of the value of the forest's biodiversity. A portion of the enterprise's revenues are invested in forest conservation and biodiversity research while conserving the Kakamega forest. The organisation is investing in mitigating the effects of climate change and enhancing community livelihoods.

The Kakamega Natural Forest Catchment Conservation Organization (KANFCCO) [26] is an NbS in the River Isiukhu watershed that is involved in environmental activities and conservation of genetic resources within the Kakamega region. The organisation promotes the growing of mukombera (*Mondia whitei*) as a mechanism to reduce the degradation of Kakamega forest for medicinal value. *Mondia whitei* is believed to be a herbal medicine with various medicinal values such as treating infertility and increasing fertility. The PELIS, Muliru Farmers Conservation Group and KANFCCO NbS programs could explain the increase in NDVI between 2005 and 2013. Furthermore, decrease in NDBSI between 1990 and 2013 could be attributed to increased vegetation cover in the region as propagated by the PELIS, Muliru Farmers Conservation Group and KANFCCO programs in the basin.

EMAUA is a Swiss-Kenyan nonprofit-making organisation that was founded in 2015 with the aim to plant 1 million indigenous trees per year by 2025 and preserve native biodiversity. EMAUA partners with community-based organisations in Kakamega and Busia counties to plant indigenous trees in protected areas such as Kakamega forest in order to improve biodiversity. Other roles include sensitising locals on the importance of protecting indigenous trees, promoting economic stability, self-reliance and sufficiency of local communities through promoting organic farming and growing of native seedlings in nurseries, encouraging use of energy saving stoves to reduce deforestation, protecting vulnerable, threatened and endangered tree species through collaborating with Kenya Forest Service and Kenya Wildlife Service.

3.3 Precipitation and Temperature Changes and Trends in the River Basin

The overall precipitation within the Isiukhu basin increased by 22.7% (427.3 mm) between 1990 and 2023 (Fig. 4). Precipitation increased by 0.44% between 1990 and 2003, 13.66% between 2003 and 2013 and 7.48% between 2013 and 2023. The general increase in rainfall can be attributed to the location of the basin in the equatorial zone as well as increased evapotranspiration due to proportional increase in vegetation cover over the same period of time.

The maximum temperature in the entire basin in 1990, 2003, 2013 and 2023 were 28.2 °C, 29.6 °C, 29 °C and 29.8 °C, while the minimum temperature in the entire basin were 12.2 °C, 13.2 °C, 12.7 °C and 13.6 °C, respectively (Fig. 4). Mean annual temperature increased by 7.43% (1.5 °C) between 1990 and 2023 (Fig. 5). This is consistent with the global warming trend [2]. Temperature increased by

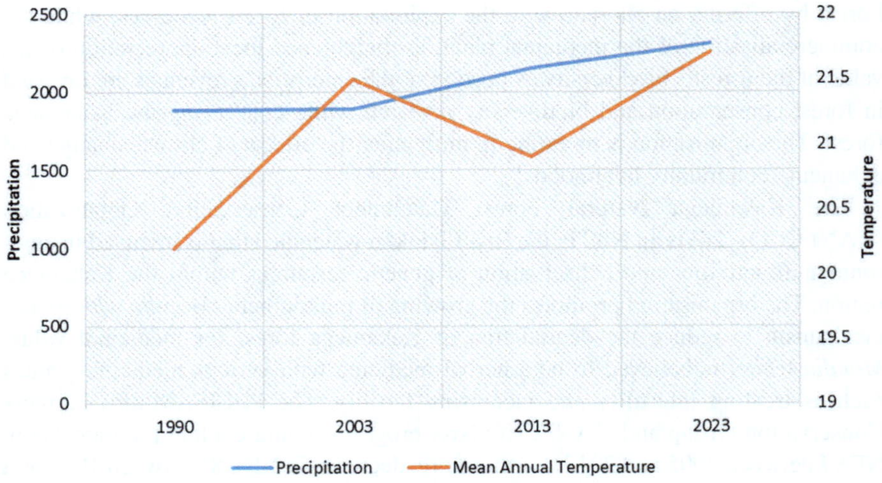

Fig. 4 Precipitation and temperature trends from 1990 to 2023

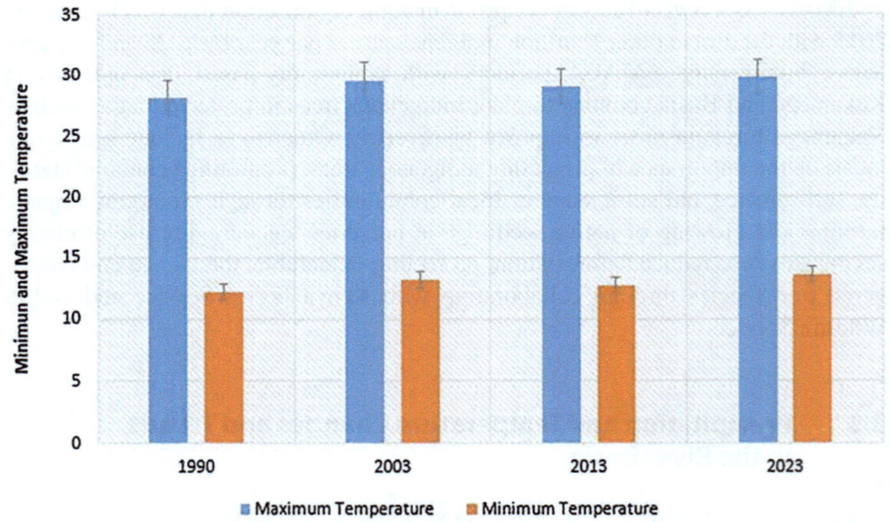

Fig. 5 Minimum and maximum temperature trends from 1990 to 2023

6.44% (1.3 °C) between 1990 and 2003, decreased by 2.8% (0.6 °C) between 2003 and 2013 and increased by 3.83% (0.8 °C) between 2013 and 2023. This shows a steady increase in temperature of the basin which is an alarming signal. The general increase in temperature over time may signify urban heat island effects as a result of increasing urbanisation within Kakamega urban centre in the basin as well as expansion of bare surface areas due to increased land degradation.

3.4 NDVI and Climatic Patterns in Relation to Nature-Based Solutions in the Study Area

The Muliru Farmers Conservation Group, Plantation Establishment and Livelihood Improvement Scheme (PELIS), Kakamega Natural Forest Catchment Conservation Organization (KANFCCO) and EMAUA are major NbS that were founded in 1997, 2005, 2012 and 2015, respectively. The figure below shows the vegetation health of the study area in relation to the establishment of the NbS in the river Isiukhu basin (Fig. 6). The study established that before the introduction of NbS projects (1990–1996), the vegetation health of the study area had an NDVI of 0.529. With the introduction of Muliru Farmers Conservation Group (1997–2004), the NDVI increased to 0.548. The NDVI value increased further to 0.55 between 2005 and 2011, a season when PELIS was introduced in the region. Between 2012 and 2014, the NDVI value increased even further to 0.6. A slight decline was recorded between 2015 and 2023 with the region recording an NDVI of 0.569. The results of this study reveal that the highest mean NDVI was achieved with the introduction of Kakamega Natural Forest Catchment Conservation Organization (KANFCCO) and EMAUA projects. This indicates that these NbS could have played a critical role in enhancement of vegetation health through afforestation and reafforestation projects in the region.

Mean seasonal precipitation increased over time with introductions of various NbS organisations and projects in the river basin. The increases were also noted in different seasons as shown in the figure below (Fig. 7). During the 1990–1996 season, the recorded mean precipitation was 463.55 mm. The amount increased to 469.29 mm for 1997–2004, 505.02 mm for 2005–2011, 565.18 mm for 2012–2014 and 568.11 mm for 2015–2023 period. These results indicate that all the NbS

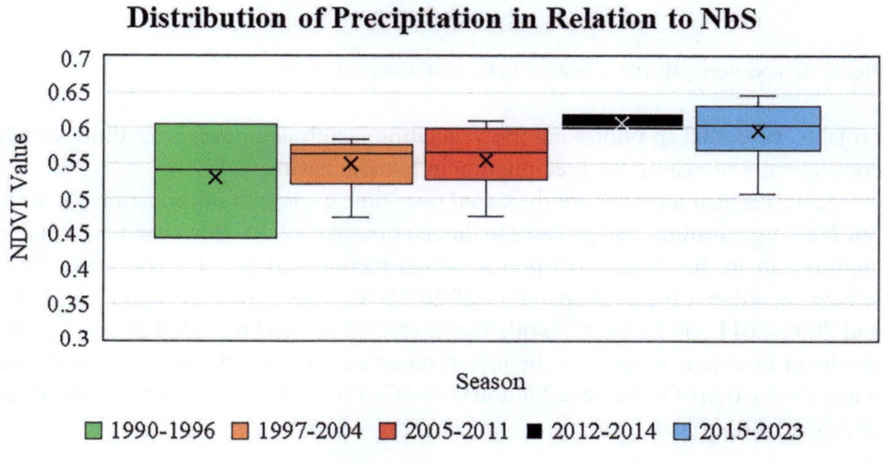

Fig. 6 Distribution of NDVI in relation to Establishment of NbS

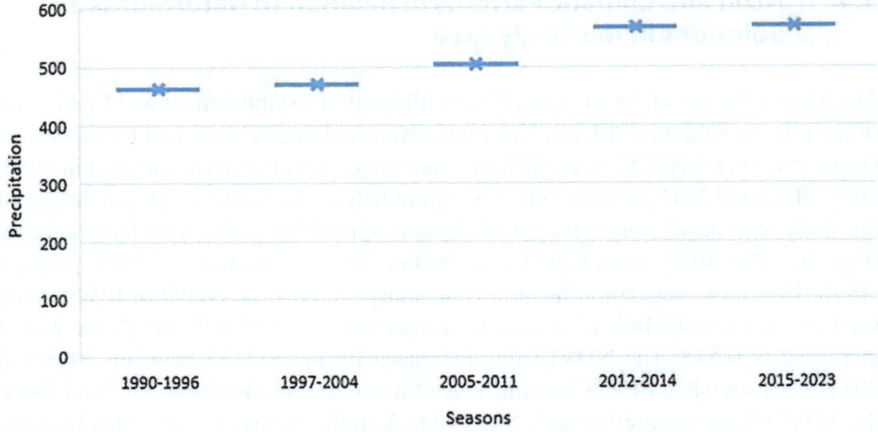

Fig. 7 Precipitation patterns in relation to Establishment of NbS

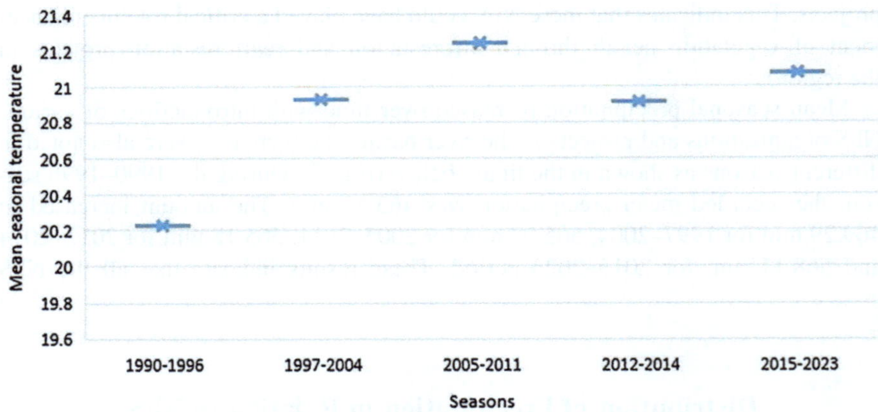

Fig. 8 Temperature patterns in relation to the establishment of NbS

projects were vital in improving the vegetation health and cover over time, which consequently increased the precipitation in Isiukhu basin.

Mean seasonal temperature increased over time even with introductions of various NbS organisations and projects in the river basin (Fig. 8). However, temperature slightly fall in the 2012–2014 period which recorded 20.91 °C. The 1990–1996 season recorded a mean temperature of 20.23 °C, 1997–2004 recorded 20.93 °C, and 2005–2011 had 21.24 °C, while the 2015–2023 period recorded 21.08 °C. The results of this study reveal that the lowest mean seasonal temperature was achieved with introduction of Kakamega Natural Forest Catchment Conservation Organization (KANFCCO) project.

3.5 Relationship Between Vegetation Health, NbS and Climate Variability

3.5.1 NDVI, NbS and Climate Variabilities

The NDVI was positively correlated with precipitation ($r = 0.5549 > 0.05$) and temperature ($r = 0.8276 > 0.05$). There was a strong positive significant correlation between NDVI and precipitation before and after introduction of various NbS ($r = 0.9692$, $p < 0.05$). The positive correlation between NDVI and precipitation signifies that vegetation health, greenness and various afforestation NbS projects in the region influenced precipitation and that the healthier the vegetation, the more rainfall due to high evapotranspiration. Similar results were noted in other studies whereby a strong temporal relationship between vegetation and rainfall occurred and that precipitation patterns were determined by vegetation phenology [2, 5]. The NbS programs adaptation such as afforestation programs and other sustainable management practices might have led to increased NDVI, consequently leading to increased rainfall in the basin [9]. There was a positive non-significant correlation between NDVI and temperature before and after introduction of various NbS ($r = 0.5097$, $p > 0.05$). The positive correlation between NDVI and temperature signifies that the ever-increasing temperature in tropical environments continues to create a favourable warm environment for the vegetative growth in a highly moist Isiukhu River basin. The findings of this study agrees that climatic variables and anthropogenic activities significantly influence vegetation cover and general ecosystem health [18, 21].

3.5.2 NDBSI and Climate Variabilities

Climatic elements varied along NDBSI gradient over 33 years. NDBSI was negatively correlated with precipitation ($r = -0.139 > 0.05$) and positively correlated with temperature ($r = 0.1916 > 0.05$). Increased NDBSI especially during 2013–2023 might have been responsible for increased temperature in the same years. Furthermore, the urban growth and the ever-expanding agronomic practises in the basin might have also increased NDBSI, which negatively affected precipitation patterns in the basin and accelerated temperature. Similar result was noted by [10], whereby increased land degradation as evidenced by increased NDBSI positively affected LST in Pakistan.

4 Conclusion

Nature-based solutions played a critical role in enhancing vegetation health of the study area, which had a direct impact on precipitation more than temperature in the study basin. The capacity of NbS to mitigate climatic changes depended on the vegetation health which is affected by natural and anthropogenic dynamics occurring on the landscape. In general, the study reported an increasing trend in NDVI and NDBSI. Similarly, the study reported an increasing trend in precipitation as well as temperature. Changes in NDVI and NDBSI influenced the temperature and precipitation patterns as depicted by the study results, thereby impacting the basin climate. Positive correlation between NDVI and precipitation and a negative correlation between NDBSI and precipitation indicated that increased vegetation health influences precipitation. A positive correlation between NDBSI and temperature indicates that land degradation increases temperature. Therefore, anthropogenic activities that improve and conserve vegetation cover should be adopted in climate change mitigation and adaptation. The policy implications of this study is to advocate nature-based solutions as a mechanism to enhance vegetation health and mitigate the effects of climate change. Therefore, the study concluded that NbS is key in mitigating the current and future climatic changes, especially precipitation patterns.

Acknowledgement We acknowledge and we are thankful to the digital resources and services provided by the United States Geology Survey, Esri Inc., Google Earth Engine, Climate Hazards Group InfraRed Precipitation with Station data (CHIRPS) and TERRACLIMATE data portal.

Authors' Contribution Samuel Abuyeka Tela conceived and designed the present study, collected and analysed the data, prepared tables, wrote the final manuscript and was the corresponding author. Nelly Nambande Masayi contributed to the present study design, materials, collected and analysed data and reviewed the draft manuscript.

Data Availability Statement The data that support the findings of this study are available on request from the corresponding author.

Conflict of Interest The authors declare that the research was conducted in the absence of any commercial or financial relationships that could be construed as a potential conflict of interest.

Funding InformationThe authors did not receive any funding.

References

1. Agevi, H., Mwendwa, K. A., Koros, H., Mulinya, C., Kawawa, R. C., Kimutai, D. K., Wabusya, M., Khanyufu, M., & Jawuoro, S. (2016). PELIS forestry programme as a strategy for increasing forest cover and improving community livelihoods: Case of Malava Forest, Western Kenya. *American Journal of Agriculture and Forestry, 4*(5), 128–135. https://doi.org/10.11648/j.ajaf.20160405.13
2. Alemayehu, Z. Y., Minale, A. S., & Legesse, S. A. (2023). Land-use and land-cover dynamics nexus to local climate variability in Suha watershed, upper Blue Nile basin, Northwest Ethiopia. *Sustainable Environment, 9*(1), 2278828.
3. Berger, C., Rosentreter, J., Voltersen, M., Baumgart, C., Schmullius, C., & Hese, S. (2017). Spatio-temporal analysis of the relationship between 2D/3D urban site characteristics and land surface temperature. *Remote Sensing of Environment, 193*, 225–243.

4. Carlson, T. N., & Ripley, D. A. (1997). On the relation between NDVI, fractional vegetation cover, and leaf area index. *Remote Sensing of Environment, 62*, 241–252. https://doi.org/10.1016/S0034-4257(97)00104-1
5. Garai, S., Khatun, M., Singh, R., Sharma. J., Pradhan, M., Ranjan., A., Rahaman, S.M., Khan, L.M., and Tiwari, S. (2022). Assessing correlation between rainfall, normalized difference vegetation index (NDVI) and land surface temperature (LST) in eastern India. Safety in Extreme Environments https://doi.org/10.1007/s42797-022-00056-2, 4, 119.
6. Gibbs, H. K., et al. (2005). Global consequences of land use. *Science, 309*, 570–574.
7. Gogoi, P. P., Vinoj, V., Swain, D., Roberts, G., Dash, J., & Tripathy, S. (2019). Land use and land cover change effect on surface temperature over Eastern India. *Scientific Reports, 9*, 8859.
8. Guha, P.M., Mburu, K., Mueller, H., Kasina, J., & Bielefeld, U (2008). Costs and benefits of conserving Kakamega forest under different three management approaches. In *Biodiversity of Africa. International Congress at Spier, RSA*.
9. Guthiga, S., & Govil, H. (2020). Land surface temperature and normalized difference vegetation index relationship: A seasonal study on a tropical city. *SN Applied Sciences, 2*, 1661.
10. Hissan, R.U., Shafiq, H., Nassani, A.A., Parveen, N., Hussain, B and Radulescu (2024). Assessment of spatial-temporal dynamics of land and desertification in Pakistan and its impact on agriculture productivity. Environment, Development and Sustainability, 26, 32337.
11. Hobbie, S. E., & Grimm, N. B. (2020). Nature-based approaches to managing climate change impacts in cities. *Philosophical Transactions of the Royal Society A, B375*, 20190124. https://doi.org/10.1098/rstb.2019.0124
12. IUNC (2016). *International Union for Conservation of nature annual report*.
13. Kumar, R., Mishra, R., Gupta, H. P., & Dutta, T. (2021). Smart sensing for agriculture: Applications. advancements, and challenges. *IEEE Consumer Electronics Magazine, 10*(4), 51–56.
14. Lato, R and Gosh, S. (2020). *Assessing the impact of spatio-temporal land cover changes on land surface temperature using satellite data in Beas Valley, Himachal Pradesh, India*.
15. Lillesand, T., Kiefer, R. W., & Chipman, J. (2015). *Remote sensing and image interpretation*. Wiley. M02 18 - 736 pages.
16. International Conference on Disaster Management and Climate Change. Earth and Environmental Science.
17. Mård, J., Box, J. E., Brown, R., Mack, M., Mernild, S. H., Walker, D., & Walsh, J. (2017). Cross-cutting scientific issues. In *Snow, water, ice and permafrost in the Arctic (SWIPA) 2017* (pp. 231–256) Arctic Monitoring and Assessment Programme (AMAP).
18. McAlpine, C. A., Syktus, J., Deo, R. C., Lawrence, P. J., McGowan, H. A., Watterson, I. G., & Phinn, S. R. (2007). Modeling the impact of historical land cover change on Australia's regional climate. *Geophysical Research Letters, 34*(22). https://doi.org/10.1029/2007GL031524
19. Meraj, G., Romshoo, S. A., Ayoub, S., & Altaf, S. (2018). Geo informatics based approach for estimating the sediment yield of the mountainous watersheds in Kashmir Himalaya India. *Geocarto International, 33*(10), 1114–1138. https://doi.org/10.1080/10106049.2017.1333536
20. Ramakreshnan, L., & Aghamohammadi, N. (2023). The application of nature-based solutions for urban Heat Island mitigation in Asia: Progress, challenges, and recommendations. *Current Environmental Health Reports, 11*, 4. https://doi.org/10.1007/s40572-023-00427-2
21. Raymondi, R. R., Cuhaciyan, J. E., Glick, P., Capalbo, S. M., Houston, L. L., Shafer, S. L., & Grah, O. (2013). Implications of changes in temperature and precipitation. *Climate Change Northwest, 41*. https://doi.org/10.5822/978-1-61091-512-0
22. Seddon, N., Daniels, E., Davis, R., Chausson, A., Harris, R., Hou-Jones, X., Huq, S., Kapos, V., Mace, G. M., Rizvi, R. A., Reid, H., Roe, D., Turner, B., & Wicander, S. (2020). Global recognition of the importance of nature-based solutions to the impacts of climate change. *Global Sustainability, 3*. https://doi.org/10.1017/sus.2020.8
23. Shipra, S., Pankaj, K., Rakhi, P., Barbaros, G., & Abhinav, R. (2024). Establishing the relationship between land use land cover, normalized difference vegetation index and land surface temperature: A case of Lower Son River Basin, India. *Geography and Sustainability, 5*, 265–275.

24. Tela. (2023). *Effects of riparian cover types on water quality and macroinvertebrate population in River Isiukhu watershed, Kakamega County, Kenya.* M.A Thesis. Kenyatta University.
25. United Nations. (2019). *World population prospects: The 2019 revision.* United Nations.
26. United Nations Development Programme. 2012. *Muliru Farmers Conservation Group, Kenya. Equator Initiative Case Study Series.*
27. Yue, W., Xu, J., Tan, W., & Xu, L. (2007). The relationship between land surface temperature and NDVI with remote sensing. Application to Shanghai Landsat 7 ETM + data. *International Journal of Remote Sensing, 28*, 3205–3226. https://doi.org/10.1080/01431160500306906

The Effect of Coumarin Additive on Optoelectronic Properties and Power Conversion Efficiency of Formamidinium-Based-Perovskite Solar Cells

Chepngetich Naomy, Mumbi Gloria, Kemboi Brendah, Tuimising Judith, Yegon K. Geoffrey, Chepkwony C. Sarah, and Koech K. Richard

Abstract

Perovskite solar cell (PSC) is among the emerging photovoltaic technologies known for their high solar power conversion efficiency (PCE) and low manufacturing costs. However, their solution-based processing leads to the formation of defects within the bulk of the perovskite layer and at its interfaces with the charge transport layers. These defects act as charge-trapping centers and provide the degradation routes which not only reduces the PCE of perovskite solar cells but also leads to instability issues. Various defect passivation strategies have been developed, but the use of additives appears to have gained preferences over the others. In this work, 1D solar cell capacitance simulator (SCAPS-1D) was employed to study the effects of coumarin additive on the PCE of formamidinium-based PSC. Parametric studies were carried out on the coumarin-modified PSC device in order to identify routes for further improvement in the PCE. It was found that a further reduction in defect density, band gap and series resistance together with an increase in the charge carrier mobilities are important in achieving higher PCE. This study provides an insight into the role of additives in enhancing the PCE of PSC, and the output serves as a reference to guide the design of effective experimental model PSC devices.

C. Naomy · M. Gloria · Y. K. Geoffrey · K. K. Richard (✉)
Department of Mathematics, Physics and Computing, Moi University, Eldoret, Kenya

K. Brendah · T. Judith · C. C. Sarah
Department of Chemistry, Moi University, Eldoret, Kenya

N. Ronoh, C. Rono (eds.), *Rethinking Our Ecosystems*, Springer Proceedings in Geography and Sustainability, https://doi.org/10.1007/978-3-031-95781-9_4

Keywords

Coumarin additive · Perovskite film · Power conversion efficiency · SCAPS-1D

1 Introduction

There is a rise in global demand for energy due to an increase in population, advancement of technology and rise in living standard. This has led to high dependency on fossil fuels which are exhaustible and can lead to global warming. Therefore, there is a need to adopt sustainable sources of energy like renewable energy. Solar energy is a preferred form of renewable energy due to the advantages of being environmentally friendly, abundant, and cheap. Among the photovoltaic technologies which deal with the conversion of solar energy directly to electricity, perovskite solar cell (PSC), ranks as one of the most attractive emerging technologies [5, 6]. This is due to its good power conversion efficiency (PCE) and low cost of fabrication. Since their inception, the PCE of PSCs has increased from 3.8% to 25.7% within a decade approaching that of conventional solar cells currently present in the market [2, 10]. This improvement is due to high absorption coefficient, long diffusion length, and tunable band gap of perovskite film [3].

The general structure of perovskite material is given by the formula ABX_3, where A is monovalent cation, B divalent cation and X is halide anion. This setup allows for extensive compositional engineering of these materials to further enhance their photovoltaic performance. Through compositional engineering, methylammonium lead iodide ($MAPbI_3$) was combined with formamidinium lead iodide ($FAPbI_3$) as a bilayer, in order to investigate the phase stability, morphology and the performance of PSC device. However, this double-cation perovskite faces challenges because of thermal instability and interfacial structure which make them sensitive to processing conditions. This has led to development of triple-cation perovskite which are made of cesium (Cs), formamidinium (FA), and methylammonium (MA) which are tenuous during the fabrication process [4]. Since the field of photovoltaic is focusing on a low cost, stable, and more efficient PSCs, additive engineering has been adopted. Additive engineering plays an important role in PSCs such as defects passivation and crystal growth [8, 9, 11].

Engineering using additives are most employed into perovskite film by mixing with precursor solutions. Additives are in two different forms; synthetic and natural additives. Synthetic additives are commonly employed; however, they are toxic to the environment and very expensive. Therefore, researchers have shifted to phytochemicals-based additives since they are environmentally friendly, cheap, and less toxic. Coumarin class of compounds is one of the widely distributed compounds in plants with potential for applications as additives in the PSC fabrication process. For example, coumarin methacrylate (CMA), which is a coumarin derivative, was used recently in CsFAMA perovskite, and due to its hydrophobic nature, the stability was improved hence increasing performance. This study sought to investigate the effects of coumarin additive on the performance of FA-based-PSC. SCAPS-1D software was employed to simulate CMA-perovskite device, in order to find the optimal values of band gap energy, defect density, and series resistance. The parametric optimization led to improved performance of PSC.

2 Methodology

In this work, solar cell capacitor simulator (SCAPS Version 3.3.11: a software developed at the department of Electronics and information system of University of Gent, Belgium by Marc Burgelman et al. [1]) was used for the optimization of some parameters like band gap energy, series resistance, and defect density of perovskite film. This software has the ability to calculate various photovoltaic performance parameters of PSCs which include but not limited to the current-voltage (I–V) characteristics, short-circuit current density (*J*sc), open circuit voltage (V_{oc}), fill factor (FF), and the power conversion efficiency. SCAPS 1D simulation software is based on solving equations of continuity and Poisson's equations for electron and hole charge carriers in a semiconductor material.

In this study, SCAPS-1D was used to simulate PSC with and without CMA additive. Some of the simulation model parameters of perovskite film were obtained from the experimental measurements of the CMA-based perovskite film from the work of Wang et al. [8]. The device structure includes fluorine-doped tin oxide, tin oxide (SnO_2), perovskite film made of $\{Cs_{0.05}(FA_{0.85}MA_{0.15})_{0.95}Pb(I_{0.85}Br_{0.15})_3\}$, Spiro-OMeTAD, and gold. The perovskite film precursor solution was modified with 0.3% of CMA as the additive. Additional parameters of this CMA-based PSC used the default values in SCAPS-1D (Table 1), and the setup of the model is shown in Fig. 1.

In this device, the band gap energy was varied between 1.55 eV and 1.63 eV and defect density from 1E+15 to 1E+16 so as to find the optimal value of both parameters in the active layer. All the simulation was performed at a temperature of 300 K under AM1.5G spectrum illumination.

The output of the simulations reported in the form of graphical presentations and optimal values of the optoelectronic properties of the CMA-based PSC was found to have better results compared to the pristine device.

3 Results and Discussion

In this section the results of simulated CsFAMA perovskite device with CMA and without were presented in the form of a J–V characteristic curve shown in Fig. 2 and it shows that the device without CMA have lower values of V_{oc}, J_{sc}, FF, and the PCE of 1.11 eV, 21.72 mA/cm2, 75.97%, and 18.48%, respectively. The CMA-based

Table 1 Input parameters of different layers used in SCAPS-1D

Input parameters	SnO_2	CsFAMA	CMA-CsFAMA	Spiro-OMeTAD
Thickness (nm)	30	490	515	200
Band gap energy (eV)	3.440	1.612	1.615	3.2
Electrical affinity (eV)	4.390	4.30	4.30	2.2
Dielectric permittivity (relative)	9.00	25.0	25.0	3.0
CB effective density of state (cm^{-3})	2.2E+18	2.2E+18	2.2E+18	2.2E+18
VB effective density of state (cm^{-3})	1.8E+19	1.3E+19	1.3E+19	1.0E+19
Electron mobility (cm^2/vs)	2.0E+1	5.1E-3	9.9E-3	1.0E-4
Hole mobility (cm^2/vs)	1.0E+1	5.0E+1	5.0E+1	2.0E-4
Shallow donor density ND (cm^{-3})	1.0E+19	1.0E+16	1.0E+16	0
Shallow acceptor density NA (cm^{-3})	0	0	0	2.8E+19
Total defect density Nt (cm^{-3})	1.0E+15	2.1E+16	6.4E+15	1.0E+15

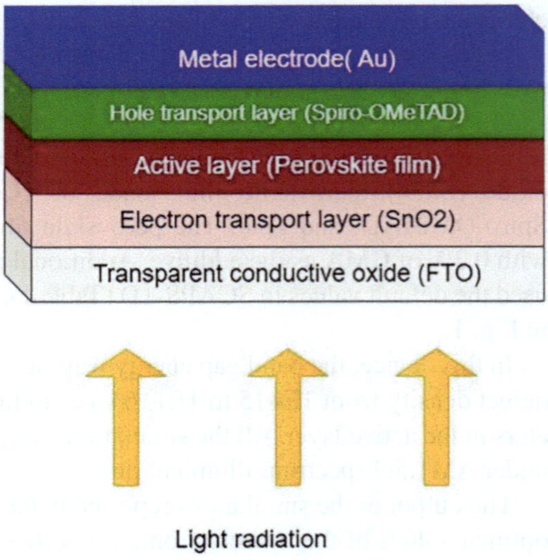

Fig. 1 Schematic diagram of simulated perovskite solar cell [7]

PSC device had a V_{oc} of 1.13 eV, J_{sc} of 21.89 mA/cm_2, FF of 80.54%, and PCE of 20.01%. CMA as an additive has a greater influence on the performance of PSC due to its hydrophobic nature and ability to act as a passivating agent at the perovskite surface hence reducing non-radiative recombination. In addition, optimization was done on some parameters like band gap, defect density, and series resistance in order to investigate the optimal value of each parameter, and thereafter the CMA-PSC device was simulated using the optimal and was seen to have better performance as shown in Table 2.

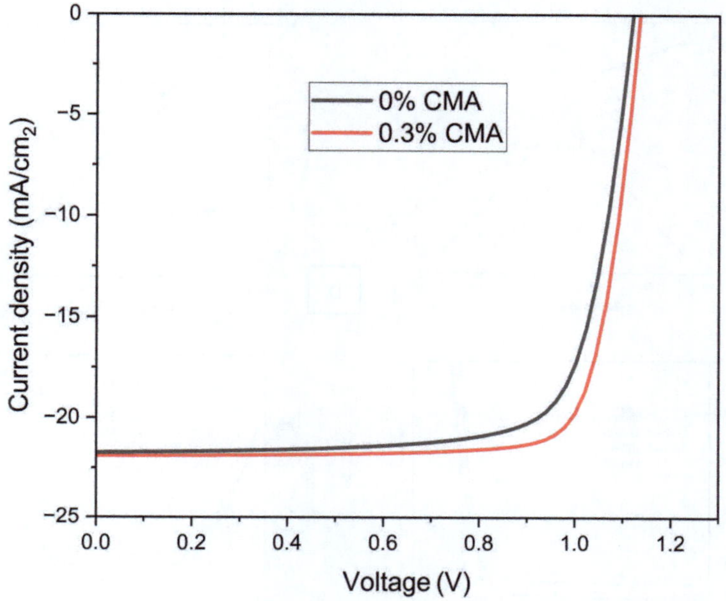

Fig. 2 I-V characteristics of perovskite device with and without CMA

Table 2 Simulated results

PSC structure	V_{oc} (V)	J_{sc} (mA/cm2)	FF (%)	PCE (%)
PSC without CMA	1.11	21.72	75.97	18.48
PSC with CMA	1.13	21.89	80.54	20.01
PSC-CMA with optimized values of Eg, Rs, and Nt	1.12	23.83	85.46	22.91

4 Effect of Band Gap

Band gap energy of the CMA-based active layer is important in determining the electrical conductivity of the solid material. Therefore, in Fig. 3a, b, the band gap values were varied from 1.55 eV to 1.63 eV. It was found that, with increase in band gap, the V_{oc} increased from 1.09 eV to 1.14 eV, while J_{sc} decreased from 23.78 mA/cm^2 to 21.50 mA/cm^2. This led to a decrease in both the fill factor from 81.68% to 80.34% and the PCE (21.25% to 19.68%) of the formamidinium-based perovskite solar cell. This means that lower values of band gap is necessary for better performance in PSCs. Therefore, mechanisms to decrease the band gap of CMA-based PSC can result in further performance enhancement. This due to increased Jsc Fig. 3c arising from better light absorption and increased recombination resistance Fig. 3d leading to better charge carrier collection as seen in Fig. 3e . The relationship between the PV parameters of a solar cell can be mathematically

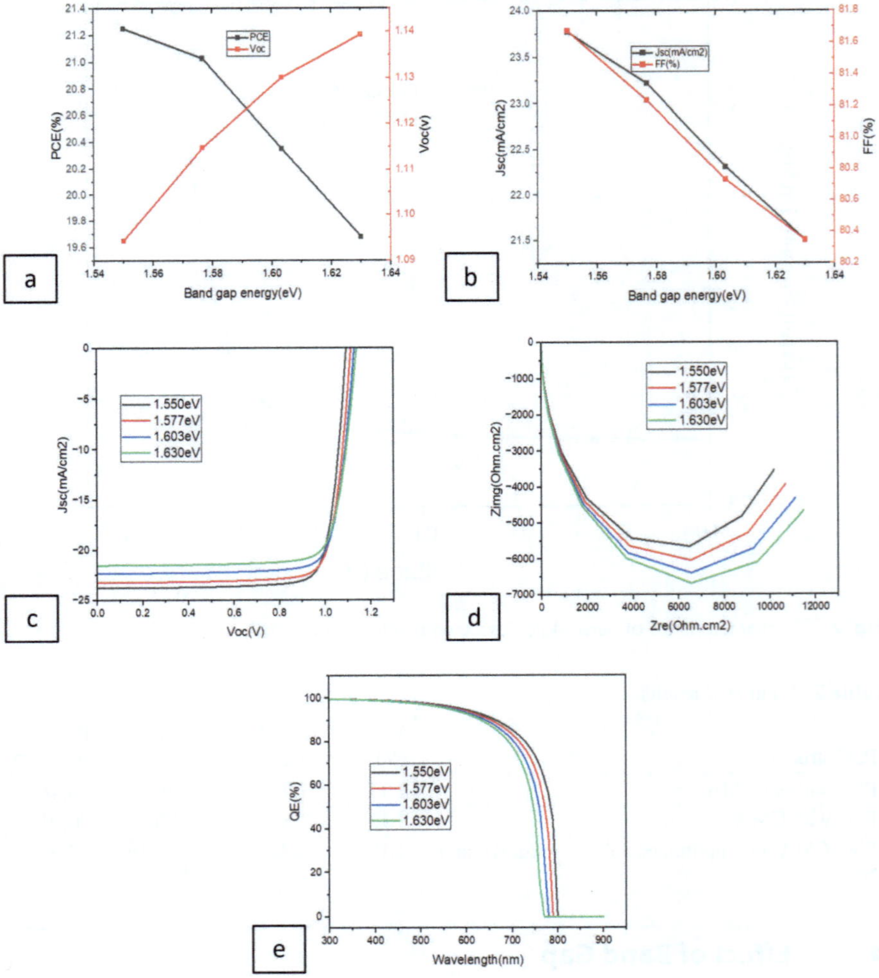

Fig. 3 (**a**) The variation of PCE and V_{oc} with band gap energy. (**b**) The variation of J_{sc} and FF with band gap energy; (**c**) I-V characteristics of varied band gap energy. (**d**) Nyquist plot of band gap energy value. (**e**) External quantum efficiency of band gap energy

expressed as follows [12]: Efforts to increase Jsc, Voc and FF will therefore lead to increased PCE in CMA-based PSCs.

$$FF = \frac{P_{max}}{J_{sc} * V_{oc}}$$

$$PCE = \frac{J_{sc} * V_{oc} * FF}{P_{in}}$$

P_{max} is the maximum power output, J_{sc} is the short circuit current density, V_{oc} is the open circuit voltage, FF is the fill factor, and P_{in} is the incident solar power.

From the impedance results (Fig. 2d), it is observed that the PSC value has a lower band gap energy of 1.55 eV and a smaller diameter compared to that of higher band gap of 1.63 eV.

5 Effect of the Series Resistance

The influence of series resistance (Rs) on the performance of CMA-PSC devices was also investigated in this study. At a thickness of 0.515 nm, Rs was changed from 0.1 ohm.cm^2 to 1.0 ohm.cm^2 in order to understand its effects on the PV parameters. Through simulations of PSC with CMA-based additive, it was observed that as the Rs decreases, the performance increases. This was done with other parameters being constant. Figure 4e shows impedance spectra in Nyquist plot of varied series resistance though it has a slight impact on the graph. It is observed that the diameter of semicircles did not change as the Rs increased from 0.1 to 1.0 Ohm.cm^2. This implies that in the range of Rs considered, the recombination resistance of CMA-PSC is not affected. In terms of external quantum efficiency, variation of Rs does not affect the EQE of the device even in the presence of additives.

6 Effects of the Defect Density

Defect density (Nt) can be found at the bulk of perovskite film or at the interface between charge transport layers and the perovskite film. This defect leads to non-recombination power loss which affects the performance of the device. The defect density of perovskite film without additive is higher than that of the perovskite film with coumarin additive since the additive acts as defect passivator. From simulations of CMA perovskite device performed using SCAPS-1D, the defect density value varied from 1E+15 cm^{-3} to 1E+16 cm^{-3}. Defect density with 1E+15 cm^{-3} gives better performance in the simulated device hence becoming the optimal value shown in Fig. 5a–b. In Fig. 5e, different values of total defect affect external quantum efficiency slightly. This means further research efforts aimed at improving the surface and bulk morphology of the CMA-based perovskite film are very necessary in order to reduce the defect density in the bulk of the perovskite layer and at the interface with adjacent layers.

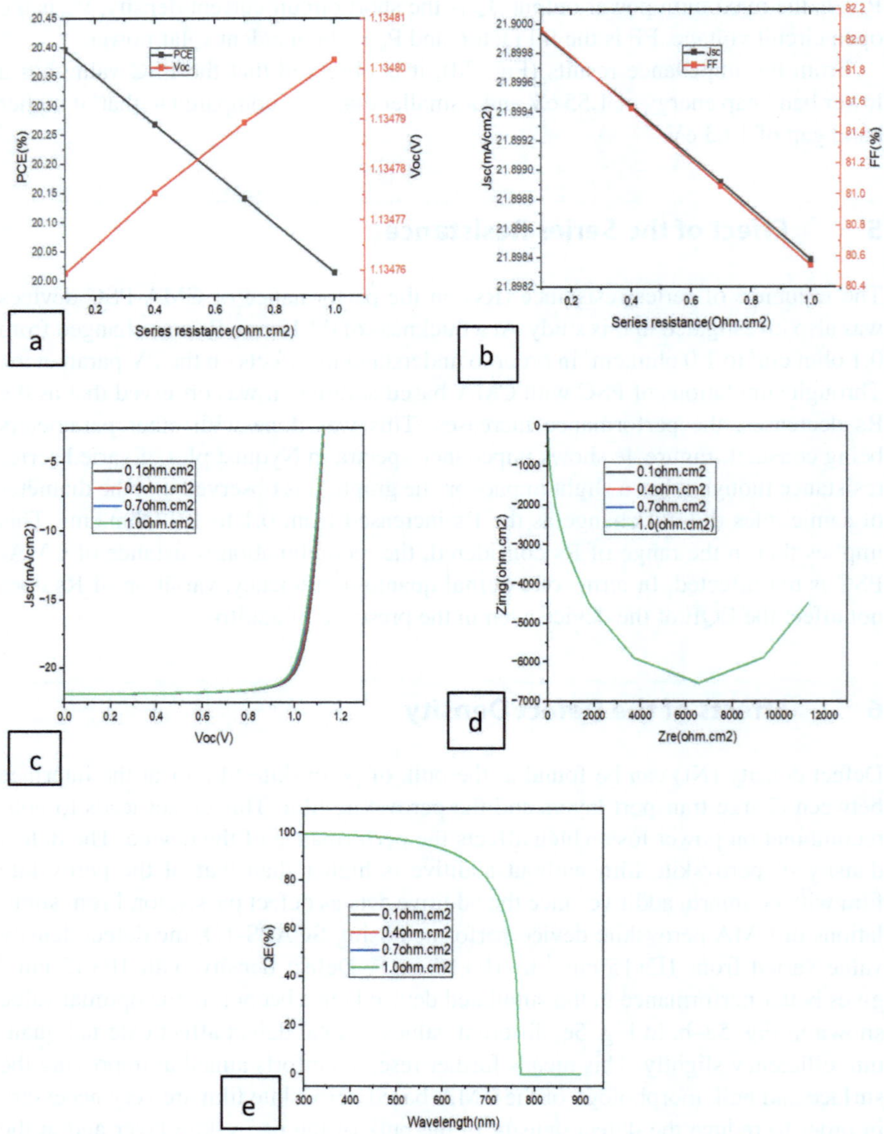

Fig. 4 (**a**, **b**) Effects of Rs of PSC on Voc, Jsc, FF, and PCE. (**c**) I-V characteristics of Rs values. (**d**) The Nyquist plot of Rs. (**e**) EQE of series resistance

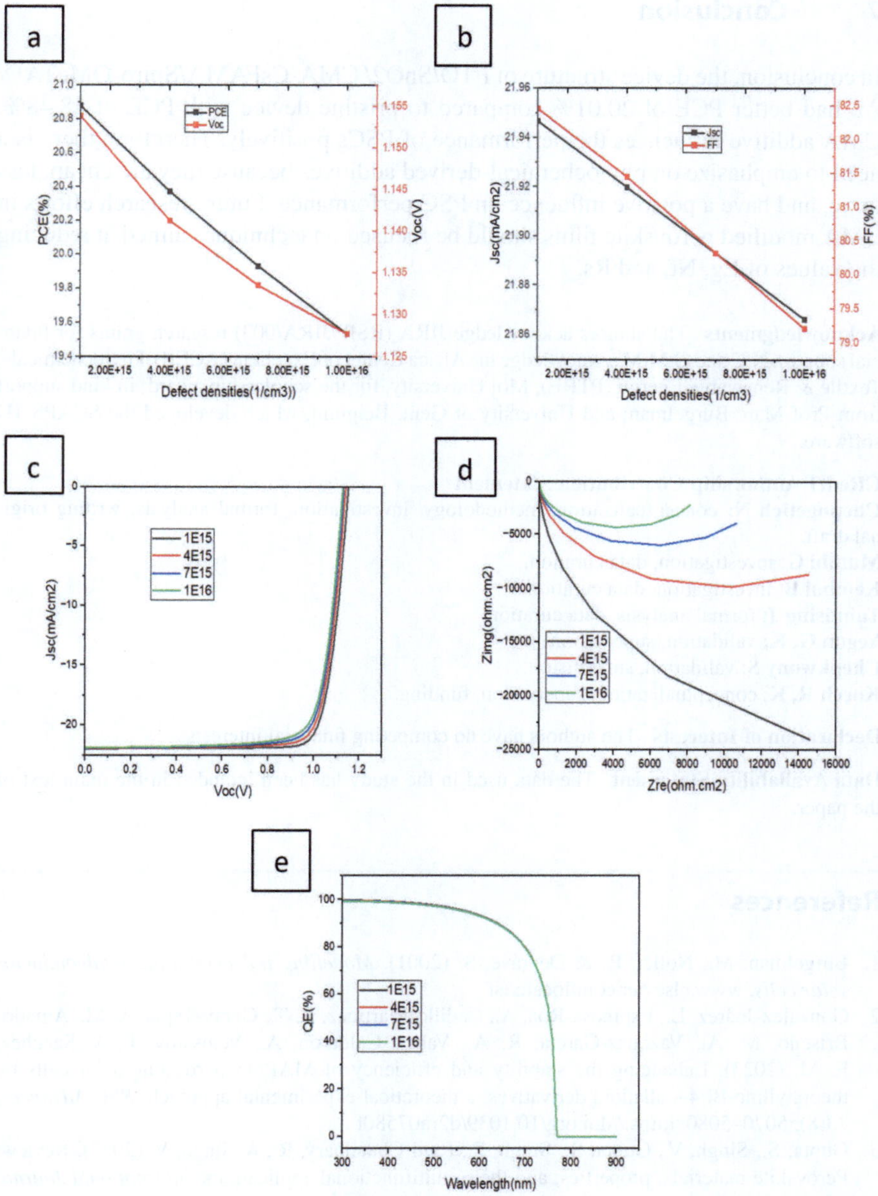

Fig. 5 (**a, b**) Effects of Nt of perovskite film on PSC performance. (**c**) Effects Nt on I-V characteristics. (**d**) Nyquist plot on varying Nt of perovskite film. (**e**) EQE graph of defect density

7 Conclusion

In conclusion, the device structure of FTO/SnO2/ CMA-CsFAMA/Spiro-OMeTAD/ Au had better PCE of 20.01% compared to pristine device with PCE of 18.48%. CMA additive influences the performance of PSCs positively. Therefore, there is a need to emphasize on phytochemical-derived additives because they are cheap, less toxic, and have a positive influence on PSC performance. Future research efforts in CMA modified perovskite films should be focused on techniques aimed at reducing the values of Eg, Nt, and Rs.

Acknowledgments The authors acknowledge JIRA (RSIF/JIRA/003) research grants for financial support. N.C and G.M.M acknowledge the Africa Center of Excellent (ACEII)-Phytochemicals, Textile & Renewable Energy (PTRE), Moi University, for the scholarship award; in kind support from Prof Marc Burgelman; and University of Gent, Belgium, which developed the SCAPS-1D software.

CRediT Authorship Contribution Statement
Chepngetich N: conceptualization, methodology, investigation, formal analysis, writing original draft.
Mumbi G: investigation, data curation.
Kemboi B: investigation, data curation.
Tuimising J: formal analysis, data curation.
Yegon G, K: validation, supervision.
Chepkwony S: validation, supervision.
Koech R, K: conceptualization, supervision, funding.

Declaration of Interests The authors have no competing financial interests.

Data Availability Statement The data used in the study has been included in the main text of the paper.

References

1. Burgelman, M., Nollet, P., & Degrave, S. (2001). *Modelling polycrystalline semiconductor solar cells*. www.elsevier.com/locate/tsf
2. González-Juárez, E., Espinosa-Roa, A., Cadillo-Martínez, A. T., Garay-Tapia, A. M., Amado-Briseño, M. A., Vázquez-García, R. A., Valdez-Calderon, A., Velusamy, J., & Sanchez, E. M. (2023). Enhancing the stability and efficiency of MAPbI3 perovskite solar cells by theophylline-BF4− alkaloid derivatives, a theoretical-experimental approach. *RSC Advances, 13*(8), 5070–5080. https://doi.org/10.1039/d2ra07580f
3. Gupta, S., Singh, V., Gupta, S., Singh, S., Rani Chaudhary, R., & Singh, V. (2022). Review: Perovskite materials, properties, and their multifunctional applications. *International Journal of Mechanical Engineering, 7*(7) https://www.researchgate.net/publication/362347600
4. Jeon, N. J., Noh, J. H., Yang, W. S., Kim, Y. C., Ryu, S., Seo, J., & Seok, S. I. (2015). Compositional engineering of perovskite materials for high-performance solar cells. *Nature, 517*(7535), 476–480. https://doi.org/10.1038/nature14133
5. Nikolaidis, P. (2023). Solar energy harnessing technologies towards de-carbonization: A systematic review of processes and systems. *Energies, 16*(17). https://doi.org/10.3390/en16176153
6. van de Ven, D. J., Capellan-Peréz, I., Arto, I., Cazcarro, I., de Castro, C., Patel, P., & Gonzalez-Eguino, M. (2021). The potential land requirements and related land use change emissions of solar energy. *Scientific Reports, 11*(1), 2907. https://doi.org/10.1038/s41598-021-82042-5

7. Wang, M., Zhao, Y., Jin, S., Lee, J.-W., Yang, Y., Jiang, X., Yin, Y., Yavuz, I., Zhu, P., Zhang, A., Han, G. S., Jung, H. S., Zhou, Y., Yang, W., & Bian, J. (2022). Rational selection of the polymeric structure for interface engineering of perovskite solar cells. *Joule, 6*(5). https://doi.org/10.1016/j.joule.2022.04.002
8. Wang, R., Altujjar, A., Zibouche, N., Wang, X., Spencer, B. F., Jia, Z., Thomas, A. G., Mokhtar, M. Z., Cai, R., Haigh, S. J., Saunders, J. M., Islam, M. S., & Saunders, B. R. (2023). Improving the efficiency and stability of perovskite solar cells using π-conjugated aromatic additives with differing hydrophobicities. *Energy and Environmental Science, 16*(6), 2646–2657. https://doi.org/10.1039/d3ee00247k
9. Xu, X., Chueh, C.-C., Yang, Z., Rajagopal, A., Xu, J., Byeok Jo, S., & Jen, A. (2017). *Ascorbic acid as an effective antioxidant additive to enhance the efficiency and stability of Pb/Sn-based binary perovskite solar cells.* http://www.elsevier.com/open-access/userlicense/1.0/
10. Kojima (https://pubs.acs.org/doi/10.1021/ja809598r)
11. Leilaeioun & Holman, 2016 (https://pubs.aip.org/aip/jap/article/120/12/123111/345364/Accuracy-of-expressions-for-the-fill-factor-of-a)

Integration of IoT and AI Technologies for Mitigating Climate Change and Restoring Ecosystems

Fancy Kiptoo, Kipngetich Godfrey, Irine C. Samoei, and Lucy Kemboi

Abstract

Climate change and ecosystem degradation are pressing environmental challenges in the twenty-first century, resulting from both natural and human activities. As global temperatures rise and natural habitats face increasing threats, innovative solutions like the Internet of Things (IoT) and artificial intelligence (AI) offer promising mitigation strategies. This study examines IoT and AI applications in climate change mitigation and adaptation, ecosystem restoration, and conservation. In climate change mitigation, the research explores how these technologies improve energy efficiency in buildings, optimize renewable energy integration, and guide sustainable transportation. For adaptation, it highlights their contributions to early warning systems for natural disasters and urban resilience. In ecosystem restoration and conservation, the study examines applications in biodiversity monitoring, habitat management, and invasive species control. The integration of these technologies enables more precise, data-driven approaches toward climate change mitigation, facilitating ecosystem restoration and endangered species protection. Findings demonstrate that IoT and AI significantly improve energy efficiency and renewable energy integration, enhance early warning systems and urban resilience, and enable more accurate monitor-

Authors Kipngetich Godfrey, Irine C. Samoei and Lucy Kemboi contributed equally to this work.

F. Kiptoo (✉) · K. Godfrey · I. C. Samoei
Department of Information and Communication Technology, Moi University, Eldoret, Kenya
e-mail: fancyjep@mu.ac.ke; godfrey_ronoh@mu.ac.ke; irenechepkoech@mu.ac.ke

L. Kemboi
Institute of Open and Distance Learning, Moi University, Eldoret, Kenya
e-mail: lucykemboi@mu.ac.ke

N. Ronoh, C. Rono (eds.), *Rethinking Our Ecosystems*, Springer Proceedings in Geography and Sustainability, https://doi.org/10.1007/978-3-031-95781-9_5

ing of biodiversity and habitats. Despite their significant potential, challenges remain in energy consumption, data management, and privacy concerns. By addressing these issues and identifying future research directions, the study provides a way forward for stakeholders to harness these technologies for a sustainable future.

Keywords
Climate change · Internet of things · Artificial intelligence · Mitigation · Ecosystem restoration · Conservation

1 Introduction

Climate change and ecosystem degradation are the environmental challenges of the twenty-first century, causing high risk to humans and nature. The Intergovernmental Panel on Climate Change (IPCC) report indicates that human activities have caused approximately 1.0 °C of global warming which is above the recommended pre-industrial levels, with far-reaching consequences resulting from natural and human systems [1]. Concurrently, the rate of ecosystem degradation and biodiversity loss has increased, threatening the stability of Earth's life-support systems [2]. In relation to these challenges, emerging technologies such as the Internet of Things (IoT) and artificial intelligence (AI) provide solutions toward addressing the negative effects of climate change. IoT refers to the network of devices connected together that has the capability to collect and exchange data, while AI refers to machines having capability of executing tasks that require human intelligence to solve day-to-day challenges [3]. The integration of IoT and AI technologies presents unique opportunities for environmental monitoring, data analysis, and decision-making in the context of mitigating climate change and ecosystem restoration [4]. This study aims to explore these opportunities by addressing the following objectives:

1. To examine IoT and AI applications towards climate change mitigation, adaptation, ecosystem restoration, and conservation
2. To assess challenges of implementing IoT and AI solutions in mitigating climate change and restoring ecosystem

By focusing on these objectives, this paper seeks to provide a systematic review of the current state of IoT and AI applications in climate change mitigation and to highlight their potential for addressing global ecological challenges on climate change mitigation.

2 Literature Review

2.1 Current State on Climate Change and Ecosystem Degradation

Recent studies indicate that the Earth's climate system is changing at an unprecedented rate. The World Meteorological Organization (WMO) reports for the last 7 years (2015–2021) have been the warmest, with the global mean temperature for the year 2021 approximately reaching 1.11 ± 0.13 °C above the recommended pre-industrial average of 1850 to 1900 [5]. The warming trend is attributed to a number of factors such as rising sea levels, increased frequency of extreme weather events, and shifts in precipitation patterns [6]. Simultaneously, ecosystem degradation continues at an alarming rate, and the Intergovernmental Science Policy Platform on Biodiversity and Ecosystem Services (IPBES) estimates that approximately 1 million animals and plant species are nowadays threatened [7]. This loss of biodiversity resulted from factors such as changes in land-use, direct exploitation of organisms, climate change, pollution, and invasive alien species [8].

3 Overview of IoT and AI Applications in Mitigating Climate Change

The application of IoT and AI in mitigating climate change has grown significantly in recent years. IoT devices are used for environmental monitoring to provide real-time data from sensors on various parameters such as air quality, soil moisture, and wildlife movements [9]. These devices range from simple sensors to more complex systems like drones and satellite-based platforms. AI, particularly machine learning algorithms, has shown great potential in analyzing the vast amounts of data generated by IoT devices. Applications include climate modeling, species identification, ecosystem mapping, and predictive analytics for natural disasters [10]. For instance, deep learning algorithms assist in analyzing satellite imagery to track deforestation and land use changes [11].

Despite the growing body of research on IoT and AI applications in mitigating climate change, there are still several gaps to be addressed, such as the integration of IoT and AI technologies presents a notable challenge, as most research has examined these technologies in isolation rather than studying their combined application in climate change contexts [12]. Another crucial limitation lies in the scalability of existing solutions, with many studies primarily focusing on small-scale or localized applications while giving insufficient attention to implementing these solutions at regional or global levels [13]. The ethical implications of widespread environmental data collection and AI-driven decision-making have not been fully explored [14], raising important questions about privacy, accountability, and social impact. Furthermore, there is a need for more interdisciplinary research that effectively combines expertise from environmental sciences, computer science, and policy studies [15], as addressing climate change requires a holistic approach that draws from multiple fields of knowledge.

4 IoT and AI Applications in Climate Change Mitigation and Adaptation

4.1 Energy Efficiency in Buildings

IoT and AI technologies play a crucial role in improving energy efficiency in buildings, which account for a significant portion of global energy consumption and greenhouse gas emissions. Smart building systems utilize IoT sensors to collect real-time data on occupancy, temperature, lighting, and energy usage. The data is then analyzed using AI algorithms to optimize energy consumption [17]. For example, the authors in [18] demonstrated that an AI-driven building energy management system reduces energy consumption by up to 30% in commercial buildings. The system uses deep reinforcement learning to optimize HVAC (heating, ventilation, and air conditioning) operations based on occupancy patterns and weather forecasts.

4.2 Optimization of Renewable Energy Integration

The integration of renewable energy sources into existing power grids presents challenges due to their intermittent nature. IoT and AI technologies can help address these challenges by enabling more accurate forecasting of energy production and demand [19]. The authors of this study have shown that a hybrid IoT-AI system could improve solar power forecasting accuracy by 25% compared to traditional methods [20]. Additionally, the system combined data from IoT weather stations with satellite imagery and used a deep learning model to predict solar irradiance.

4.3 Sustainable Transportation

In the transportation sector, IoT and AI were used to develop smart traffic management systems that helped to optimize route planning and therefore, lowering emissions. For example, the authors of the paper [21] developed an AI-powered traffic signal control system that reduced the average vehicle delay by 20% and $CO2$ emissions by 12% in a simulated urban environment.

4.4 Early Warning Systems for Natural Disasters

IoT sensor networks combined with AI predictive models can significantly improve early warning systems for natural disasters such as floods, wildfires, and earthquakes. In study [22], researchers developed an IoT-based flood early warning system that could predict flood events with 93% accuracy up to 72 hours in advance.

4.5 Urban Resilience

Smart city initiatives leveraging IoT and AI technologies are enhancing urban resilience to climate change impacts. For example, authors of this paper developed an AI-driven urban heat island mitigation system that uses IoT sensors to monitor urban microclimates and optimize the deployment of green infrastructure, resulting in a 2 °C reduction in surface temperatures in pilot areas [23].

5 IOT and AI Application in Restoring Ecosystems and Conservation

5.1 Biodiversity Monitoring

IoT devices and AI algorithms are revolutionizing biodiversity monitoring by enabling automated species identification and population tracking. The study used camera traps and deep learning algorithms to automatically identify and count wildlife, achieving 96.8% accuracy across 48 species [24].

5.2 Habitat Management

IoT sensors are used to monitor environmental conditions in various ecosystems, providing valuable data for habitat management. For instance, in this paper authors developed an IoT-based system for monitoring mangrove forests, which used salinity, water level, and temperature sensors to provide real-time data on ecosystem health [25].

5.3 Invasive Species Control

AI-powered early detection systems and IoT-enabled targeted treatment approaches are enhancing efforts to control invasive species. Research has shown that AI models can identify invasive plant species from drone imagery with 95% accuracy, enabling more efficient and targeted eradication efforts [26].

5.4 Endangered Species Protection

IoT and AI technologies are being employed in anti-poaching efforts and wildlife corridor management to protect endangered species. The study developed an AI system that uses data from GPS collars and camera traps to predict elephant movements and potential human-elephant conflict areas, improving the effectiveness of conservation efforts [27].

6 Methodology

6.1 Systematic Literature Review Approach

This study employed a systematic literature review approach, focusing mainly on the peer-reviewed journals, conference proceedings, and reports from recognized institutions, published between 2019 and 2024. The primary databases explored included Web of Science, Scopus, IEEE, Research Gate, and Google Scholar. The goal of the review was to assess the current advancements in IoT applications in climate change mitigation and ecosystem restoration, with a particular emphasis on identifying key trends and challenges in IoT and AI technologies for mitigating climate change. Additionally, the research incorporated case study analysis of real-world projects involving IoT and AI implementations for environmental applications. This provided valuable insights into practical experiences, highlighting both the effectiveness and challenges faced in applying these technologies to mitigate climate change impacts and restore ecosystems.

6.2 Search Criteria

The literature search was conducted across key academic databases, including Web of Science, Scopus, IEEE Xplore, Research Gate, and Google Scholar. These databases were chosen for their extensive coverage of peer-reviewed articles, technical reports, and conference proceedings.

6.3 Search Strings

Search queries were developed directly from the research objectives, incorporating Boolean operators such as AND, OR, and NOT were applied to assist us to refine the search results. The primary search string combined key concepts: ("Internet of Things" OR "IoT") AND ("artificial intelligence" OR "AI") AND ("climate change" OR "ecosystem restoration" OR "environmental conservation") AND (mitigate* OR restor* OR adapt*).

6.4 Inclusion Criteria

Studies were included if they focused on the integrated application of IoT and AI in environmental contexts, provided measurable outcomes or real-world case studies, and finally, if the papers addressed key challenges or outlined future directions in these fields.

6.5 Exclusion Criteria

The studies reviewed excluded the papers published date was before 2019, focused exclusively on either IoT or AI without integration, and lacked sufficient data or peer-reviewed evidence.

6.6 Existing Procedures and Workflow

The systematic literature review of the study process narrowed down by following the PRISMA (Preferred Reporting Items for Systematic Reviews and Meta-Analyses) framework to enhance transparency and reproducibility. The search criteria followed steps started with identification, screening of the title and abstract using inclusion and exclusion criteria, assessing eligibility of the full text, and finally inclusion stage considered for analysis.

7 Results and Discussion

This study extracted authors' information on selected papers based on the type of IoT devices, AI algorithms used, and environmental application areas focusing on mitigating climate change and restoring ecosystems. We also focused on the scale of implementation such as local, regional, or global, and finally the system benefit outcomes and limitations of IoT and AI on climate change mitigation and ecosystem restoration as summarized on Table 1.

Our survey reveals that the integration of IoT and AI technologies has significant potential for addressing climate change mitigation and ecosystem restoration. Compared to earlier studies, recent research shows a trend toward more integrated and large-scale applications of IoT and AI in climate change contexts. There is also an increasing focus on real-world implementations and quantitative assessments of effectiveness, moving beyond proof-of-concept studies [28]. Some of the key findings including the following: energy efficiency improvements of 20–30% is realized in buildings through smart management systems [18]; enhanced renewable energy integration through improved forecasting accuracy [20]; reduction in urban transportation emissions by 10–15% through AI-optimized traffic management [21]; and finally, improved accuracy for natural disaster predictions [22] and effective biodiversity monitoring and conservation efforts [24, 27].

The findings suggest that IoT and AI technologies can significantly enhance our ability to monitor, understand, and manage environmental systems. However, successful implementation requires interdisciplinary collaboration and careful consideration of local contexts [29].

Table 1 Analysis of related work on IoT and AI toward climate change mitigation and ecosystem restoration (source: authors' own contribution)

Author's	Type of IoT devices	AI algorithms	IoT and AI application areas	Scale of implementation	Reported system outcome	Challenges
Zhang et al. (2020) [20]	Solar monitoring sensor	Hybrid prediction AI models	Solar prediction AI models	Regional	Enhanced solar power prediction	Prediction accuracy limitations
Jones et al. (2022) [26]	Drone imagery	AI detection	Invasive species monitoring	Local/Regional	Enhanced species identification	Image processing challenge
Liu et al. (2022) [22]	Water level sensor, sensor networks,	Deep Learning	Flood prediction	Regional	Advanced prediction models	Sensor network reliability
Jiang et al. (2023) [21]	Urban traffic sensor	AI powered optimization	Urban emissions reduction	Urban	Reduced transportation emissions	Complex urban infrastructure
Chen et al. (2023) [12]	Urban Sensor networks	Integrated IoT-AI systems	Urban carbon monitoring	Urban	Detailed carbon emission tracking	Data privacy issues
Zhang et al. (2023) [17]	IoT building sensors (smart meters, temperature sensors	Clustering, Reinforcement learning	Smart building energy optimization	Local/Urban	Improved energy efficiency	Complex building systems, retrofitting challenges
Alamgir et al. (2023) [25]	IoT Environmental sensors	Integrated monitoring systems	Conservation of Mangrove	Local	Improved ecosystem monitoring	Environmental sensor durability
Tabak et al. (2023) [24]	IoT camera networks, GPS trackers	Deep learning	Wildlife monitoring	Regional	Automated species tracking	Technological complexity

8 Study Findings and Future Work

Despite the potential benefits that come with integration of IoT and AI technologies in climate change mitigation and ecosystem restoration, there are also some of the challenges that need to be addressed such as energy consumption of IoT devices and the widespread deployment of IoT devices raises concerns about their energy consumption. While these devices aim to reduce overall energy use, their own power requirements could be significantly high [30]. Our systematic review both aligns with and extends beyond previous review studies in several key aspects. While Guo et al. (2023) and Chen et al. (2023) established fundamental frameworks for IoT-AI integration, our review synthesizes recent evidence of their effectiveness across multiple implementation contexts. Our analysis of challenges in energy consumption and data management builds upon Wang et al.'s (2023) work on scaling issues, while providing a more detailed examination of potential solutions suggested in recent literature. Our findings suggests several key gaps in current research, particularly in areas such as the integration of edge computing capabilities for climate change mitigation, the role of explainable AI in environmental decision-making, the potential of blockchain technology in climate change monitoring systems, and finally building strong interdisciplinary research and collaboration among different experts from different fields of specialization.

9 Conclusion

The integration of IoT and AI technologies offers significant potential for mitigating climate change and restoring ecosystems. Our study review highlights numerous successful applications of integrating IoT and AI on climate change mitigation and ecosystem restoration, focusing more on energy efficiency and renewable energy integration to biodiversity monitoring and conservation. However, realizing the full potential of these technologies requires addressing challenges related to energy consumption, data management, and technical integration. Future research should focus on developing more efficient and accessible IoT-AI systems, exploring novel applications, and ensuring that these technologies contribute to equitable and sustainable climate change management. As we face the unpredicted climate change, the thoughtful and strategic implementation of IoT and AI technologies can play a crucial role in our efforts toward the realization of a more sustainable future. Therefore, policymakers, researchers, and practitioners must collaborate to harness these technologies effectively while addressing associated ethical and societal implications.

Author's Contributions Conceptualization and methodology: Fancy Kiptoo, Kipngetich Godfrey, Irine C. Samoei, and Lucy Kemboi.

Conflict of Interest Statement The authors declare no conflicts of interest.

Funding InformationThis research received no external funding.

References

1. IPCC. (2023). *Climate change 2023: Synthesis report. Contribution of working groups I, II and III to the sixth assessment report of the intergovernmental panel on climate change.*
2. Díaz, S., et al. (2022). *The global assessment report on biodiversity and ecosystem services.* Intergovernmental Science-Policy Platform on Biodiversity and Ecosystem Services (IPBES).
3. Salam, A. (2022). *Internet of things for environmental sustainability and climate change.* Springer Nature.
4. Guo, H., et al. (2023). Artificial intelligence and IoT for environmental sustainability: A comprehensive review and future directions. *Environmental Research, 206*, 112609.
5. World Meteorological Organization. (2023). *State of the global climate 2022.*
6. Trenberth, K. E., et al. (2023). Climate change and extreme events. *Nature Climate Change, 13*(1), 4–6.
7. IPBES. (2022). *Global assessment report on biodiversity and ecosystem services.*
8. Pörtner, H. O., et al. (2023). *Scientific outcome of the IPBES-IPCC co-sponsored workshop on biodiversity and climate change.* IPBES Secretariat.
9. Ferrero, R., et al. (2023). IoT-based environmental monitoring: A comprehensive review of current trends and future directions. *IEEE Internet of Things Journal, 10*(5), 4531–4559.
10. Rolnick, D., et al. (2022). Tackling climate change with machine learning. *ACM Computing Surveys, 55*(2), 1–96.
11. Pham, T. D., et al. (2023). Deep learning for large-scale satellite-based mapping of deforestation and forest degradation. *Remote Sensing of Environment, 280*, 113215.
12. Chen, Y., et al. (2023). Integrated IoT-AI systems for environmental monitoring: A systematic review. *Environmental Modelling & Software, 158.*
13. Wang, Z., et al. (2023). Scaling challenges in environmental IoT deployments: A systematic review. *Environmental Monitoring and Assessment, 195*(2), 1–18.
14. Kumar, S., & Smith, R. (2023). Ethical considerations in environmental AI: Privacy, bias, and social impact. *AI & SOCIETY, 38*(4), 1089–1104.
15. Martinez, R., et al. (2022). Interdisciplinary approaches to environmental monitoring: Combining social and technical perspectives. *Environmental Science & Policy, 127*, 98–110.
16. Zhang, L., et al. (2023). Smart building energy management: Integration of IoT and AI for sustainable operations. *Energy and Buildings, 277*, 112673.
17. Wang, Y., et al. (2023). Deep reinforcement learning for building energy optimization: A case study of commercial buildings. *Applied Energy, 331*, 120301.
18. Liu, H., et al. (2023). AI-driven renewable energy integration: Challenges and opportunities. *Renewable and Sustainable Energy Reviews, 179*, 113021.
19. Zhang, X., et al. (2023). Enhanced solar power forecasting using hybrid IoT-AI systems. *Solar Energy, 251*, 114–126.
20. Jiang, L., et al. (2023). AI-powered traffic management for reduced emissions: A smart city approach. *Transportation Research Part D: Transport and Environment, 116*, 103544.
21. Liu, J., et al. (2022). Advanced flood prediction using IoT sensor networks and deep learning. *Water Research, 213*, 118162.
22. Martínez-España, R., et al. (2023). Urban heat mitigation through smart infrastructure: An IoT-AI integrated approach. *Sustainable Cities and Society, 89*, 104368.
23. Tabak, M., et al. (2023). Automated wildlife monitoring using deep learning and IoT camera networks. *Methods in Ecology and Evolution, 14*(5), 1159–1171.
24. Alamgir, M., et al. (2023). IoT-based monitoring system for mangrove ecosystem conservation. *Ecological Indicators, 146*, 109721.
25. Jones, K., et al. (2022). AI-powered detection of invasive plant species using drone imagery. *Ecological Informatics, 71*, 101942.

26. Dufernauer, R., et al. (2024). Predictive modeling of elephant movements using IoT tracking and machine learning. *Conservation Biology, 38*(1), e14088.
27. Park, J., et al. (2023). Evolution of environmental monitoring: From traditional methods to IoT-AI integration. *Environmental Monitoring and Assessment, 195*(3), 338.
28. Wilson, A., et al. (2023). Implementation challenges of environmental IoT-AI systems: A global perspective. *Environmental Science & Technology, 57*(15), 5789–5801.
29. Brown, M., et al. (2023). Energy consumption patterns in environmental IoT networks. *IEEE Internet of Things Journal, 10*(7), 6251–6265.
30. Thompson, S., et al. (2023). Data governance frameworks for environmental monitoring systems. *Environmental Data Science, 2*, e13.

Adaptation Strategies

Farmers' Adaptation to Climate Change in Nakuru County, Kenya

Leonard Emarete Karani, Rose C. Ramkat, Alfred Serem, Philip Chemwok, and Damien Jourdain

Abstract

As a key sector in Kenya's economy, agriculture is threatened by the changing climate over the years, which has negatively influenced productivity. As a result, farmer adaptation methods are now widely acknowledged as essential and vital responses to the threat posed by present and future climate changes. The objectives of this study were to examine farmers' perceptions of current and anticipated climate change and the consequences on their livelihood activities, identify both current and potential future adaptation strategies adopted in response to climate changes, and analyze the factors influencing farmers' selection of these strategies. A total of 2 focus group discussions (FGDs) and 240 structured interviews within Subukia sub-County were carried out. The findings of this study highlight increased climate variability, with hot temperatures and strong sun cre-

Authors Rose C. Ramkat, Alfred Serem, Philip Chemwok and Damien Jourdain contributed equally to this work.

L. E. Karani (✉) · A. Serem · P. Chemwok
Department of Agricultural Economics and Resource Management, Moi University, Eldoret, Kenya

R. C. Ramkat
Department of Biological Sciences, Moi University, Eldoret, Kenya

D. Jourdain
Department of Agricultural Economics, Extension and Rural Development, University of Pretoria, Pretoria, South Africa

CIRAD, UMR G-EAU, Hatfield, South Africa

G-EAU, Universite' de Montpellier, AgroParisTech, CIRAD, IRD, IRSTEA, Montepellier, France

N. Ronoh, C. Rono (eds.), *Rethinking Our Ecosystems*, Springer Proceedings in Geography and Sustainability, https://doi.org/10.1007/978-3-031-95781-9_6

ating significant stress for farmers in the study area. Key adaptation strategies employed by farmers included early planting, livestock rearing, planting short-season crops, tree planting, irrigation, and engaging in off-farm activities. The study also found that factors such as group membership, remittances, access to irrigation, and access to climate information, education level, and primary income source influenced the choice of adaptation strategies. Furthermore, land size and group membership were key factors influencing the number of new livelihood activities adopted by farmers. In conclusion, fostering group membership, improving access to climate information, expanding irrigation systems, and promoting tree planting are essential steps to strengthen farmers' resilience to climate change.

Keywords
Agriculture · Climate change · Adaptation strategies · Kenya

1 Introduction

Climate change constitutes a significant worldwide concern of our period, imposing a substantial strain on both societies and the environment [1]. Moreover, it has been identified as a hindrance to key economic sectors that rely heavily on natural resources, including agriculture, fisheries, livestock, and tourism [2]. According to [3], the ongoing climate changes present a serious threat to global food security, potentially undermining agriculture's ability to sustain the rural poor and other vulnerable populations. Furthermore, the consequences of climate change exacerbate issues related to food security and run the danger of undoing progress gained in the fight against hunger and malnutrition, especially in the most vulnerable countries and populations [4].

Driven by human-caused greenhouse gas (GHG) emissions, global climate change is increasing the sensitivity of agricultural production systems [5]. Rising global temperatures have intensified severe weather phenomena including severe rain, floods, frost, droughts, and heat waves, significantly affecting agricultural conditions and the entire crop production value chain [6, 7]. The two climatic parameters of economic and social importance are temperature and rainfall, with rainfall being of particular concern [8]. Rising temperatures and decreasing precipitation have been identified as key contributors to declining agricultural productivity and increase in the frequency of pests and diseases [9].

Africa as a continent is facing serious repercussions of climate change, with the agricultural sector being the most affected. According to the Intergovernmental Panel on Climate Change, crop production might fall by up to 50% and crop revenue by up to 90% by 2100 [10]. Agriculture remains the primary economic activity sustaining millions of livelihoods across the continent [11]. Furthermore, research suggests that the continent's temperature is currently warmer than it was a century ago, and model-based forecasts predict further warming, with the possibility of acceleration owing to greenhouse gas (GHG) emissions [12]. Additionally, rising temperatures and frequent droughts are said to particularly affected smallholder farmers in the continent [13].

Compared to other regions of the world, sub-Saharan Africa (SSA) stands out as the most affected by the adverse effects of climate change [14]. Further, the region is expected to face serious consequences of climate change because it currently experiences increasing temperatures and fluctuating rainfall [12]. Studies indicate that countries in SSA face significant challenges in responding to changes in climate due to a lack of adaptation ability [15]. Farmers in this nations are among the hardest hit by ongoing climate shifts [16]. According to the IPCC (2021), the region is predicted to demonstrate higher level of exposure to variability in climate by the end of the century coupled with decreased productivity from farming [17].

In Kenya, incidences of water scarcity, drought, flooding, and occurrence of water-related diseases characterize changes in climate [1]. Like other regions worldwide, climate change in Kenya has altered weather patterns and seasonal cycles, placing stress on agricultural ecosystems and jeopardizing agricultural productivity [18, 19]. To mitigate these impacts and enhance agricultural growth, the Kenyan government has established a number of programs. Notably, the National Climate Change Response Strategy (NCCRS) [20] and the Kenya Climate Smart Agriculture Strategy (KCSAS) 2017–2026 [21] were implemented to address climate change-related challenges.

Nakuru County is already experiencing the effects of climate change, with noticeable variations in climatic conditions over the years. Drought, heavy rains, floods, and rising temperatures are straining agricultural productivity, incomes, and food security, with these challenges expected to intensify in both frequency and severity [22]. Climate change has also been linked to unpredictable and declining crop yields, as well as the emergence of new crop and livestock diseases. Further, seasonal patterns have shifted, with rainfall becoming increasingly erratic and unreliable [22]. Projections for 2021–2065 indicate that prolonged moisture stress will persist throughout both growing seasons, while intense precipitation is expected to remain largely unchanged [22].

Farmers across the world utilize diverse adaptation measures to respond to negative effects of climate change [23]. Adaptation in agriculture can be a viable solution to mitigate against adverse impacts of climate change [24]. Further, adaptation in agriculture is vital and necessary for sustainability of livelihoods in developing countries [25]. Typical strategies for adapting to changing conditions include introducing improved crop and livestock breeds, practicing intercropping, utilizing irrigation and water conservation methods, and diversifying both within and outside agriculture. Other approaches involve integrating crop and livestock farming, adjusting planting schedules, using crop insurance, relying on early warning systems, migrating for better opportunities, and liquidating assets when necessary [26, 27, 28]. Despite ongoing efforts, the climate is already changing, and there is a widespread agreement that future climatic conditions will diverge significantly from the present, with more extreme outcomes expected [29]. Consequently, farmers might be compelled to adjust their livelihood activities in response to these changing climate conditions, highlighting the importance of examining potential adaptation strategies for these future (and more extreme) scenarios.

This study adopts a contingent behavior (CB) approach inspired by the work of [30] to explore possible adaptation strategies to future climate changes. The CB approach, like the contingent valuation method (CVM), involves asking individuals to express their preferences regarding environmental resources or changes in resource status by responding to hypothetical scenarios [31]. The approach employs well-structured hypothetical scenarios that outline both present conditions and possible future changes to directly capture individuals' behavioral responses to environmental shifts [32, 33]. According to [30], this method addresses the limitations of previous climate change adaptation studies by incorporating the variability of future climatic conditions in to the study design.

Prompted by unfolding climate uncertainty and its adverse effects on their nature-based livelihoods, farmers' have been adopting various strategies to cushion themselves from adverse effects of climate change. These measures are also evolving as climatic conditions change. Some of these strategies may provide a foundation for future climate adaptation but have yet to be fully explored. This study aims to (1) examine farmers' perceptions of current and anticipated climate change and the consequences on their livelihood activities, (2) identify both current and potential future adaptation strategies adopted in response to climate changes, and (3) analyze the factors influencing farmers' selection of these strategies.

2 Materials and Methods

2.1 Description of the Study Area

The study was undertaken in Nakuru County; specifically, the wards of Subukia, Kabazi, and Waseges were surveyed. Subukia and Kabazi fall under zone 3 of the climatological zones of Nakuru County, characterized by annual rainfalls ranging from 1100 to 1400 mm. The wards lie within an altitude of between 1800 and 2300 m above sea level. The areas are very suitable for agricultural activities [34]. Waseges falls on the dry side of the county and is characterized by rainfall ranging from 800 to 1100 mm annually. The ward lies on a general elevation of between 900 m and 1800 m above sea level and is characterized by few agricultural activities. The main agricultural activities in the study area include crop farming: maize, beans, vegetables, Irish potatoes, tomatoes, coffee, tea, and livestock farming and horticultural production. Trade also takes place and involves the sale of agricultural produce such as cereals, horticultural produce, and livestock.

2.2 Research Design

A research design provides a structured framework for carrying out a study to answer specific research questions [35]. This study employed a cross-sectional survey design to explore perceptions with regard to climate change and potential responsive strategies and examine the factors influencing the selection of the strategies among farmers in Nakuru County. This approach entails collecting data at a single point in time [36]. The design has been adopted in many social science studies due to its accuracy and precision [37]. Additionally, this design is suitable as it accommodates both quantitative and qualitative data.

2.3 Sampling Procedure and Sample Size

A multistage sampling technique was employed to select participants for this study. In the initial stage, Nakuru was deliberately chosen as the study area due to its significant agricultural activity and its susceptibility to the impacts of climate change. In the second stage, administrative wards of Subukia, Kabazi, and Waseges were purposively selected due to their accessibility and the presence of many smallholder farmers who practice rain-fed cropping. In the final stage, systematic sampling was applied whereby an interval of five households after each survey was applied to gather information from the smallholder farmers in these areas. To determine the sample size, formula by [38] was utilized:

$$n = \frac{Z^2 p(1-p)}{e^2} \tag{1}$$

where

n = sample size
Z = z value (1.96 for 95% confidence interval)
P = estimated proportion of population (assumed to be 80% or 0.8)
e = desired margin of error (assumed to be 0.05)

Therefore using the formula, we obtain a sample size of 256 respondents.

$$n = \frac{1.96 * 1.96 * 0.8(1-0.8)}{(0.05)^2} = 246 \tag{2}$$

2.4 Data Collection

Data collection was conducted in two stages: (a) the first stage involved focus group discussions (FGDs), and (b) the second stage involved household surveys.

2.4.1 Focus Group Discussions

In the first stage, two focus group discussions (FGDs) were held in the study area: one in Tetu, Subukia Ward, and the other in Polepole, Waseges Ward. Each FGD included 15–21 participants, primarily consisting of farmers and agricultural officers. The purpose of these FGDs was to gather farmers' perceptions of current and future changes in terms of climate and to identify both existing and potential climate change adaptation strategies in the area. More specifically, we wanted to better understand (1) what are the important current and anticipated climate changes, (2) how would it translate in terms of impact on their farms, (3) what are the major changes made by farmers in recent years (and in response to what?), and (4) what are the possible adaptation strategies they could use. Adaptation strategies were then ranked accordingly from the highest possible to the least one.

2.4.2 Household Surveys

Household surveys were carried out using a semi-structured questionnaire, programmed in Kobo Toolbox data collection software. The questionnaire included both open-ended and closed-ended questions. The closed-ended questions aimed to gather data that could be easily analyzed and classified in a specific manner, while the open-ended questions allowed for the collection of qualitative insights, opinions, and detailed explanations, offering a deeper understanding of the respondents' viewpoints. Structured interviews were conducted face-to-face with either the household head (male or female) at their residence, or with an adult family member in the absence of the household head. Data on socio-economic factors, institutional aspects, climate change scenarios, and adaptation strategies were collected. To construct climate change scenarios, we focused on defining future climate conditions in a way that is relevant and meaningful to the local context, with an emphasis on potential impacts on local behavior. As part of the contingent behavior (CB) approach, one specific climate change scenario was developed: heat waves (characterized by high temperatures and intense sunlight). Climate data indicated that the region already experiences at least 3 months of heat waves annually, and projections suggest that these heat waves could become more intense, extending from 13 weeks with 4.1 days to as much as 42.4 days (Table 1).

2.5 Data Analysis

2.5.1 Descriptive Analysis

Descriptive statistics, such as tabulations, cross-tabulations, frequency tables, and general measures like means and standard deviations of specific variables, were calculated. The data collected in this study were analyzed using Microsoft Excel, STATA version 15, and R programming software.

2.5.2 Empirical Model Specification

The purpose of the study was to evaluate farmers' adaptation to climate change in Nakuru County. We focused on two key aspects of household responses to future climate change: (1) the likelihood of adopting specific livelihood activities and (2) the number of activities adopted. In the first aspect, we used a logit model to determine the probability of adopting a particular activity or strategy, considering one strategy at a time. The logit model is specified as follows:

Table 1 Climate change scenario included in the contingent behavior questions

	Mild increase	Moderate increase	Extreme increase
Heat waves	Increase in heat waves by 2 weeks	Increase in heat waves by 4 weeks	Increase in heat waves by 6 weeks

Source: Author's (2025)

$$Y_i^* = \alpha + X_i\beta + \varepsilon_i$$
$$Y_i = \{1\, if\, Y_i^* > 0\, 0\, otherwise \tag{3}$$

where Y_i is an observed variable that indicates whether a farmer adopts a particular strategy ($Y_i = 1$), if $Y_i^* > 0$, and zero otherwise. The X_i represents determinants that affect the choice of an adaptation strategy through coefficients, β. The errors, ε, are assumed to have a standard logistic distribution.

Since farmers were able to tick several strategies, the multinomial model was not adequate. However, we tested whether there exist possible correlations in the adoption of the different strategies using the multivariate probit models. Following the works of [39], a system of simultaneous probit models is specified as follows:

$$Y_{im}^* = \beta_m^{'} X_{im} + \varepsilon_{im}$$
$$Y_{im} = \{1\, if\, Y_{im}^* > 0\, 0\, otherwise \tag{3.4}$$

The error terms, ε_{im}, and $m = 1, \ldots, M$ are distributed as multivariate normal, each with a mean of zero and a variance–covariance matrix v, where V has a value of 1 on the leading diagonal and correlations ρjk = ρkj off-diagonal elements.

In the second case, we explained the number of new alternative strategies chosen by the respondents. Given that the variable is a count variable, we employed a Poisson model to analyze the factors affecting the choice of adaptation strategies by an individual. The Poisson model is specified as follows:

$$Prob(X_i) = \frac{\exp(-\lambda_i)\lambda_i^{Y_i}}{Y_i!}, \quad \lambda_i = expexp(X_i\beta), \quad Y_i = 0, 1, \ldots, n, \tag{3.5}$$

where Y denoted the number of adaptation strategies chosen in response i, X_i represents the explanatory variables, and β, s are the parameters to be estimated. The Poisson regression model has a log-linear conditional mean function, $\lambda(i.e., E(X_i) = \lambda_i)$ [40].

The independent variables that were used in the empirical model analysis together with their definitions and hypothesized signs are presented in Table 2.

3 Results and Discussions

3.1 Farmers Perceptions of Current and Future Climate Changes

This subsection presents the findings from the two focus group discussions (FGDs) held in Tetu, Subukia Ward, and Polepole, Waseges Ward. The results highlight

Table 2 Variable measurements and a priori expectations

Variable	Measurement	Expected sign
Dependent		
Adaptation strategy choice	1 = yes, 0 = no	
Number of new livelihood activities	In numbers	
Independent		
Socio-economic		
Age	In years	+
Gender of the respondent	1 = male, 0 = female	+/−
Gender of the household head	1 = male, 0 = female	+/−
Education level	1 = none, 2 = primary, 3 = secondary 4 = tertiary	+
Farming experience	In years	+
Household size	In numbers	+
Distance to input and output market	In kilometers	+/−
Income source	1 = farm-based, 2 = monetary income, 3 = off-farm, 4 = pension, 5 = remittances	+/−
Remittances	1 = yes, 0 = no	+
Land size	In acres	+
Irrigation	1 = yes, 0 = no	+
Institutional		
Extension visits	1 = public, 2 = private, 3 = no	+
Credit access	1 = yes, 0 = no	+
Group membership	1 = yes, 0 = no	+
Access to climate information	1 = yes, 0 = no	+

Source: Author's (2025)

participants' perceptions of weather pattern changes in recent years, the adaptation strategies they have employed, and the expected future changes along with the likely adaptation measures they anticipate.

3.1.1 Results on the First Community: Subukia-Tetu

The focus group discussions revealed that participants in the study area observed several climatic events, including high temperatures, untimely rains, long dry spells, little rain, strong winds, and long rains. Farmers were successively requested to rank the climatic events in terms of current frequency (from most frequent: 6 to least frequent: 1), impact on their livelihoods (from strong negative impact: 6 to weak impact: 1), vulnerability (from strong vulnerability: 6 to small vulnerability: 1), and future occurrence (from most frequent: 6 to least frequent: 1). Table 3 of the Borda count results on the frequency of occurrence, impact, vulnerability, and future occurrence revealed that untimely rains were the most frequent, while strong winds

Table 3 Borda scores for frequency, impact, vulnerability, and future occurrence

Event	Frequency	Impact	Vulnerability	Future occurrence
High temperatures	75	88	86	84
Untimely rains	112	50	69	101
Long dry spell	80	95	77	68
Little rain	70	77	63	93
Strong winds	43	65	68	49
Long rains	61	66	78	46

Source: Author's (2025)

were the least frequent climatic scenarios reported by participants. In addition, the results on the impact of various climatic events on livelihood activities indicated that long dry spells had the greatest impact, whereas untimely rains had the least. Furthermore, the results on vulnerability demonstrated that farmers were most vulnerable to hot temperatures and least vulnerable to little rain. Finally, regarding the likelihood of future occurrence, the results suggested that untimely rains are the most likely to occur, while long rains are the least likely.

Table 4 displays the composite scores results generated to account for different types of problems created by different climatic events. The scores were calculated as the geometric mean of the scaled scores obtained for future frequency, impact, and vulnerability. Each scores were first scaled using a min-max normalization of the scores in order to make them commensurable. Two composite scores were compared: $Fr \times Imp \times Vuln$ used the scores of future frequency, impact, and vulnerability; $Fr \times Imp$ used the scores of future frequency and impact only. The results reveal that hot temperature and long dry spell created more problems as compared to the other climatic events.

The focus group discussions identified a range of adaptation strategies employed by participants in the study area. These included irrigation, mixed farming, organic farming, early planting, chemical use, crop diversification, crop rotation, and value addition, among others. Farmers were asked to rank these strategies, with six indicating the most commonly used and one representing the least commonly used, for both current and anticipated strategies. Table 5 of Borda count results on the adaptation strategies revealed that early planting was ranked as the most used adaptation strategies, while planting high value crops was the least used. The results also indicated that, for future adaptation strategies, participants ranked mixed and organic farming as the most likely strategies to be adopted, while value addition was considered the least likely.

3.1.2 Results in the Second Community: Waseges-Polepole

Focus group discussions (FGDs) conducted in this area revealed low rains, strong sun, unpredictable rain, and frost as the climatic change events of interest to the participants. Farmers were successively requested to rank the climatic events in terms of current frequency (from most frequent: 6 to least frequent: 1), impact on their livelihoods (from strong negative impact: 6 to weak impact: 1), vulnerability (from strong vulnerability: 6 to small vulnerability: 1), and future occurrence (from

Table 4 Scaled Borda and composite scores

Event	Frequency	Future frequency	Impact	Vulnerability	Fr × Imp × Vuln	Fr × Imp
Untimely rains	0.999	0.998	0.002	0.257	0.083	0.047
Long dry spell	0.535	0.398	0.998	0.604	0.622	0.630
Hot temperatures	0.462	0.689	0.842	0.996	0.833	0.762
Little rain	0.390	0.853	0.598	0.004	0.130	0.714
Long rains	0.259	0.002	0.353	0.648	0.075	0.025
Strong winds	0.001	0.053	0.331	0.213	0.155	0.132
1 Borda scores have been min-max scaled so they fit in [0,1] intervals						

Source: Author's (2025)

Table 5 Borda scores for adaptation strategies in Subukia Ward-Tetu

Adaptation strategies	Current	Rank	Future	Rank
Irrigation	32	5	33	4
Mixed farming	50	3	54	1
Organic farming	57	2	54	1
Early planting	59	1	53	3
Chemical use	7	8	2	10
Change of crops	26	6	31	6
Crop rotation	47	4	33	4
Value addition	4	10	1	11
Short season crops	26	6	31	6
Planting high-value crops	2	11	3	9
Switch to other enterprises	5	9	19	8

Source: Author's (2025)

most frequent: 6 to least frequent: 1). Table 6 of Borda count results on the frequency of occurrence ranked strong sun as the most frequent climatic event, followed by low rains, and then unpredictable rain, with frost as the least frequent climatic event. Additionally, the results assessing the impact on livelihood activities ranked strong sun as the climatic event that most affected participants, followed by low rains and then frost, with unpredictable rain having the least impact on participants. Moreover, the results based on vulnerability indicated that participants felt most vulnerable to strong sun, followed by frost, then low rains, and were least vulnerable to unpredictable rains. Finally, regarding the likelihood of future occurrence, the participants ranked unpredictable rain as the most likely climatic event, followed by low rains and then strong sun, with frost as the least likely event.

Table 7 displays the composite score results generated to account for different types of problems created by different climatic events. The scores were calculated as the geometric mean of the scaled scores obtained for future frequency, impact, and vulnerability. Each scores were first scaled using a min-max normalization of the scores in order to make them commensurable. Two composite scores were

Table 6 Borda scores on frequency, impact, vulnerability, and future occurrence

Event	Frequency	Impact	Vulnerability	Future occurrence
Low rains	43	44	30	43
Strong sun	45	51	57	37
Unpredictable rain	42	32	28	47
Frost	30	33	45	33

Source: Author's (2025)

Table 7 Scaled Borda and composite scores

Event	Frequency	Future frequency	Impact	Vulnerability	Fr × Imp × Vuln	Fr × Imp
Strong sun	0.993	0.279	0.995	0.997	0.651	0.526
Low rains	0.860	0.707	0.626	0.066	0.307	0.666
Unpredictable rain	0.793	0.993	0.005	0.003	0.026	0.072
Frost	0.007	0.007	0.047	0.583	0.058	0.018
1 Borda scores have been min-max scaled so they fit in [0,1] intervals						

Source: Author's (2025)

Table 8 Borda scores for adaptation strategies in Waseges Ward-Polepole

Adaptation strategies	Current	Rank	Future	Rank
Intercropping	49	2	32	4
Early planting	73	1	58	1
Crop rotation	12	7	9	7
Livestock rearing	31	3	39	2
Plant early maturing crops	26	5	32	4
Plant trees	28	4	32	4
Chemical use	19	6	35	3
Change of seeds	2	8	3	8

Source: Author's (2025)

compared: *Fr × Imp × Vuln* used the scores of future frequency, impact, and vulnerability; *Fr × Imp* used the scores of future frequency and impact only. The results reveal that strong sun and low rains created more problems as compared to the other climatic events.

Table 8 displays Borda count results with regard to the adaptation strategies showed that early planting was ranked top while change of seeds was ranked last.

3.2 Descriptive Statistics Results for Farmers

Table 9 presents results for the socio-economic characteristics and institutional factors of small-scale farmers in the study area. The table of results showed that 53.33% of the survey participants were female, and 46.67% were male. This indicates a

Table 9 Socio-economic and institutional characteristics of farmers in Nakuru County

Variable	Description	Frequency	Percentage (%)
Gender of respondent	Male	128	53.33
	Female	112	46.67
Gender of household head	Male	59	24.58
	Female	181	75.42
Education level	Nonformal	14	5.83
	Not finished primary school	37	15.42
	Finished primary school	70	29.17
	Not finished secondary	29	12.08
	Finished secondary	64	26.67
	Not finished college/university	8	3.33
	Finished college/university	18	7.50
Experience	Less than 5 years	15	6.25
	5 to 10 years	41	17.08
	More than 10 years	184	76.67
Income source	Farm-based	176	73.33
	Monetary income	22	9.17
	Off-farm	37	15.42
	Pension	1	0.42
	Remittances	4	1.67
Remittances	No	171	71.25
	Yes	69	28.75
Irrigation access	No	186	77.50
	Yes	54	22.50
Extension visits	No	159	66.25
	Private	27	11.25
	Public	42	17.50
	Public and private	12	5.00
Credit access	No	210	87.50
	Yes	30	12.50
Group membership	No	192	80.00
	Yes	48	20.00
Information access	No	131	54.58
	Yes	109	45.42

Source: Author's (2025)

small gender difference in the survey participation, with females being the majority. Further, this shows that women were highly involved in farming activities and easily available for surveys as compared to their male counterparts.

The table of results also reveals that 75.42% of the households were headed by males, while 24.58% were headed by females. This indicates that a significant majority of the households are led by males, whereas a smaller proportion are led by females. This suggests a notable gender disparity in household leadership and decision-making, with men predominantly in the role of household head. In terms of education level, 29.17% of small-scale farmers in the study area finished primary school, 26.67% finished secondary level education, 15.42% did not complete primary school, 12.08% did not complete secondary education, 7.5% finished college or university, 5.83% had no formal education, and 3.33% never finished college or university. The findings indicate a diverse range of educational attainment among the respondents, the majority of having completed primary-level education.

Additionally, the results table reveals that 76.32% of farmers in the study area have been farming for more than 10 years, 17.37% have less than 5 years of experience, and 6.25% have been farming for 5–10 years. This indicates that most farmers in the area possess substantial farming experience. Further, it translates to the acquisition of skills and knowledge gained through sustainable involvement in agricultural activities. The results further show that 73.33% of the farmers' primary source of income is farming, 9.17% comes from monetary sources, 15.42% from off-farm activities, 0.42% from pension schemes, and 1.67% from remittances. This indicates that the majority of farmers in the study area rely primarily on farming for their income.

The results table shows that 71.25% of farmers in the study area did not receive any remittances in 2023, while 21.75% did. This suggests that the majority of farmers in the area do not receive external financial support through remittances. Additionally, the findings reveal that 77.5% of farmers did not irrigate any of their land in the previous year, while 22.5% did irrigate part of their land. This indicates that most farmers in the study area did not adopt irrigation practices in 2023.

The results table shows that in 2023, 66.25% of farmers did not receive visits from extension service providers, 17.5% were visited by public extension service providers, 11.25% by private extension service providers, and 5% by both public and private providers. This indicates that many farmers lacked direct contact with extension services in 2023. Additionally, the findings reveal that 87.5% of farmers did not obtain credit for farming in 2023, while 12.5% did. This suggests that most farmers did not have access to credit for their farming activities and were not reliant on credit farming.

The results table also shows that 80% of farmers were not members of any farmer group, while 20% were members of various farmer groups. This suggests a low level of participation in such groups, potentially limiting many farmers' access to benefits like collective action, knowledge sharing, and resource availability. Additionally, the findings regarding access to climate-related information revealed that 54.58% of farmers did not receive any climate-related information, while 45.42% obtained it from different sources. This highlights a potential gap in access to important climate-related information among farmers.

Table 10 presents the summary statistics for the continuous variables used in the study, which include the respondent's age, household size, distance to the nearest market, land owned by the farmer, land rented, and land under cultivation. The results indicate that the average age of the respondents in the study area is 49.87 years, with the youngest respondent being 6 years old and the oldest being 96 years old.

According to the results table, the average household size in the study area was 3 members, with smaller households consisting of just 1 person and larger

Table 10 Descriptive statistics for continuous variables

Variable	Obs	Mean	Std. dev.	Min	Max
Age (years)	240	48.571	16.536	6	96
Household size	240	3.821	1.972	1	14
Distance to market (kms)	240	2.554	2.433	0.5	25
Land owned (acres)	240	1.826	1.669	0	15

Source: Author's (2025)

households having up to 14 members. The table also shows that, on average, farmers traveled about 2.554 km to reach the nearest market, with the closest market being 0.5 km away and the farthest market located 25 km from the household. Additionally, the results revealed that the average land owned by farmers in the study area was 1.826 acres, with the smallest plot being 0.125 acres and the largest at 15 acres.

3.3 Current and Future Climate Change Adaptation Strategies

Table 11 presents the percentage of responses reflecting current livelihood activities and future adaptation strategies in light of the proposed climate change scenario. The results show that 52.92% of farmers practice early planting, while 48.33% of farmers currently practice livestock farming. Further, it is evident that 84.58% of farmers currently plant short season crops, 53.75% tree planting, and 27.08% practice irrigation. Additionally, results show that 36.25% of farmers engage in off-farm activities, and 0.83% of farmers do not practice the stated adaptation strategies. Moreover, the findings reveal a high preference for short-season crops and off-farm activities in response to heat waves.

3.4 Factors Influencing the Choice of Adaptation Strategies to Future Climate Changes

Table 12 displays the logit regression results regarding the factors influencing the choice of adaptation strategies in response to different durations of heat waves, which were incorporated as future climate change scenarios in this study. The results table on factors influencing the adoption of early planting as an adaptation strategy indicates that, out of the 14 variables analyzed, only 3 were found to be statistically significant. These variables—main income source, group membership, and extension visits—were positively associated with the adoption of early planting.

Group membership was found to have a positive and statistically significant influence on the decision to adopt early planting, with a significance level of 5%. Specifically, farmers who were members of farmer groups had an 89.7% higher likelihood of adopting early planting compared to nonmembers. The increased likelihood could be due to the support, resources, shared knowledge, or motivation

Table 11 Percentage of response showing current livelihood activities and future adaptation strategies under different climatic scenario

	Climate			
Adaptation Strategies	Current	Increase by 2 weeks	Increase by 4 weeks	Increase by 6 weeks
	%	%	%	%
Early planting	52.92	20.90	21.15	28.99
Livestock rearing	48.33	22.39	26.92	21.74
Short season crops	84.58	52.24	50.96	47.83
Tree planting	53.75	19.40	19.23	23.19
Irrigation	27.08	34.33	29.81	31.88
Off-farm activities	36.25	32.84	21.15	40.58
None of the above	0.83	16.42	15.38	14.49

Source: Author's (2025)

provided within the group, which encourages timely decision-making aligned with early planting practices. These findings are consistent with those of [16], who found that membership in a farmer group positively influenced the adoption of climate change adaptation strategies in Ghana. Similarly, [11] observed that affiliation with groups enhanced the likelihood of introducing new breeds in response to climate variability in Kenya.

Remittances were shown to have a positive and statistically significant effect, at the 5% significance level, on the likelihood of adopting short-season crops. This suggests that households receiving remittances were 73.7% more likely to opt for short-season crops. This suggests that households receiving remittances are much more likely to invest in short-season crops, likely due to the added financial flexibility that remittances provide. The extra income from remittances enables farmers to manage the risks or costs associated with planting short-season crops, which are often employed to counter the negative effect of changing environmental conditions. The present results are comparable with those of [41], who also showed a favorable association between remittances and the adoption of adaption techniques among pastoral communities in Kenya.

Irrigation access was discovered to have a favorable and statistically significant effect on the decision to grow short-season crops at the 5% level. This indicates that having undertaken irrigation activity in 2023 increases the chances of selecting short-season crops by 88.6%. Irrigation allows farmers to manage water supply more effectively, making it easier to grow these crops, especially in areas experiencing variable rainfall or drought conditions. Further, irrigation reduces the risks associated with short-season crops, making them a more viable option for farmers.

Access to climate-related information was found to be favorable and statistically significant at the 1% level of significance in influencing the decision to plant trees. Informed farmers are well equipped to understand climate trends, such as shifting weather patterns and environmental risks, prompting them to adopt tree planting to mitigate climate change impacts. The current findings are similar with those of [23], who observed that access to climate information increased the probability of

Table 12 Logit regression results on the factors influencing the choice of an adaptation strategy

	Early planting	Livestock rearing	Short season crops	Tree planting	Irrigation	Off-farm activities	None of the above
Explanatory variable	Coef. (P-value)	Coef. (P-value)	Coef. (P-value)	Coef. (P-value)	Coef. (P-value)	Coef. (P-value)	Coef. (P-value)
Age	0.016 (0.213)	−0.013 (0.255)	**−0.018 (0.095)***	0.008 (0.551)	0.011 (0.302)	−0.019 (0.108)	**0.034 (0.03)****
Gender of Respondent	0.375 (0.348)	0.127 (0.723)	−0.189 (0.57)	−0.214 (0.587)	−0.041 (0.909)	0.345 (0.354)	**−0.979 (0.04)****
Gender of household head	−0.341 (0.459)	0.385 (0.392)	−0.098 (0.806)	0.317 (0.501)	−0.157 (0.702)	0.486 (0.298)	0.607 (0.264)
Education level	0.089 (0.311)	0.025 (0.761)	**−0.136 (0.07)***	−0.121 (0.202)	0.036 (0.658)	**0.232 (0.005)*****	−0.099 (0.365)
Household size	0.071 (0.466)	−0.041 (0.63)	0.008 (0.913)	−0.009 (0.924)	−0.014 (0.872)	**−0.148 (0.093)***	0.017 (0.887)
Remittances	−0.204 (0.614)	−0.077 (0.84)	**0.737 (0.035)****	−0.105 (0.798)	−0.157 (0.665)	**−1.047 (0.017)****	−0.125 (0.802)
Income source	**0.365 (0.059)***	−0.062 (0.749)	0.006 (0.972)	0.188 (0.308)	−0.303 (0.101)	**0.582 (0.001)*****	0.195 (0.391)
Distance to market	−0.077 (0.431)	0.007 (0.912)	−0.008 (0.893)	0.017 (0.796)	0.003 (0.962)	−0.006 (0.931)	0.018 (0.85)
Experience	−0.184 (0.473)	0.173 (0.462)	0.061 (0.77)	0.277 (0.327)	0.351 (0.167)	0.169 (0.459)	**−0.467 (0.096)***
Land owned	0.145 (0.149)	−0.005 (0.963)	0.044 (0.659)	0.049 (0.641)	−0.08 (0.383)	0.086 (0.398)	−0.196 (0.249)
Irrigation access	0.59 (0.122)	−0.02 (0.957)	**0.886 (0.014)****	−0.003 (0.995)	**1.042 (0.003)*****	−0.239 (0.555)	−0.551 (0.319)
Group membership	**0.897 (0.033)****	0.352 (0.396)	0.493 (0.216)	−0.016 (0.971)	−0.146 (0.717)	−0.432 (0.372)	0.142 (0.79)
Credit access	−0.481 (0.354)	0.198 (0.667)	0.217 (0.624)	−0.138 (0.793)	0.556 (0.221)	0.025 (0.96)	−0.448 (0.518)
Information access	0.539 (0.125)	0.171 (0.596)	**0.53 (0.069)***	**1.091 (0.003)*****	**1.106 (0.001)*****	−0.156 (0.64)	**−1.148 (0.011)****
Extension visits	**0.17 (0.084)***	0.075 (0.415)	**−0.169 (0.056)***	0.055 (0.581)	−0.016 (0.858)	−0.015 (0.877)	0.044 (0.722)
Constant	−4.941 (0.002)	−2.649 (0.075)	−1.697 (0.189)	−4.161 (0.012)	−4.622 (0.002)	−0.797 (0.594)	1.014 (0.561)
Mean dependent var	0.233	0.242	0.504	0.204	0.317	0.300	0.154

	Early planting	Livestock rearing	Short season crops	Tree planting	Irrigation	Off-farm activities	None of the above
Explanatory variable	Coef. (P-value)	Coef. (P-value)	Coef. (P-value)	Coef. (P-value)	Coef. (P-value)	Coef. (P-value)	Coef. (P-value)
SD dependent var	0.424	0.429	0.501	0.404	0.466	0.459	0.362
Pseudo r-squared	0.121	0.023	0.093	0.075	0.118	0.143	0.101
Chi-square	31.485	6.074	30.799	18.296	35.297	41.890	20.939
Prob > chi^2	0.008	0.978	0.009	0.247	0.002	0.000	0.139
Number of obs	240	240	240	240	240	240	240

*** $p < .01$, ** $p < .05$, and * $p < .1$ denotes significance at 1%, 5%,and 10%, respectively
Source: Author's (2025)

planting food and fodder trees as an adaptation strategy in Ethiopia. Similarly, [42] highlighted that information motivated farmers to invest in adaptation techniques in Ethiopia.

Irrigation access was associated with an increased likelihood of selecting irrigation as a responsive measure to changing climatic conditions. This suggests that having engaged in any form of irrigation activities in 2023 increases the odds of adopting irrigation by 104.2%. This implies that once farmers decide to begin irrigating, the probability of fully adopting irrigation as a regular practice rises significantly, perhaps due to increased awareness, resources, or infrastructure put in place during the period. The current findings are convergent with those of [43], who observed that having irrigated land increased the probability of practicing irrigation as a climate-smart technology in Ethiopia.

Farmers who had access to climate-related information were significantly more likely to choose irrigation. Specifically, this access increased the likelihood of selecting irrigation by a substantial 110.6%, a statistically significant finding at the 1% level. This suggest that farmers who are informed about climatic conditions, such as rainfall patterns or drought risk, are significantly more likely to implement irrigation systems that improve water management and adapt to climate variability. This access to information also allows farmers to make informed and sustainable decisions about their agricultural practices. Our findings are consistent with [44]'s report of a positive influence of climate information on adaptation adoption in Ethiopia, but contradict [45]'s finding of a negative association between these factors in the same country.

Education level significantly increased the likelihood of choosing off-farm activities at the 1% significance level. The probability of engaging in off-farm activities is likely to increase by 23.2% with one-level increase in education. Individuals with higher education are more inclined to pursue opportunities beyond farming, likely due to enhanced skills, knowledge, or increased access to alternative employment options. The current findings are convergent with those of [46], who observed that education increased the possibility of utilizing various measures in India. Similarly, a study in rural Tanzania by [47] revealed a significant increase in the uptake of responsive measures due to increase level of education. However, the results differ from those of [48] who found that decision to embrace livelihood diversification option among farmers reduced with education.

Income source of the farmer significantly increased the likelihood of engaging in off-farm activities at 1% significance level. Specifically, having a stable or diverse income source increases the odds of partaking in off-farm activities by 58.2%. Farmers with more reliable income streams are better positioned to explore additional livelihood options beyond agriculture, as they have greater financial flexibility and access to resources needed for off-farm ventures. The current findings corroborate with those of [49], who found out that income sources positively influence the selection of livelihood adaptation strategies in Pakistan. Likewise, [38]'s research in Kenya indicated a positive correlation between the amount of income generated from farming and pastoralists' decisions to implement climate change adaptation strategies.

Remittances were found to have a negative and statistically significant effect at the 5% significance level on the decision to undertake off-farm activities. Specifically, receiving remittances decreases the odds of participating in off-farm activities by 104.7%. As an income source, remittances can directly enhance household welfare [50]. This substantial negative marginal effect implies that households receiving remittances are significantly less likely to invest in off-farm income generating activities. This might be because remittances provide sufficient support, reducing the need for individuals to seek additional income from other income-generating activities.

An increase in age is positively associated with an increase in the likelihood of not undertaking a specific adaptation strategy. This implies that with each additional year in a farmer's age, the probability of not engaging in the specified adaptation strategies increases by about 3.4%. As farmers grow older, their accumulated experience may lead them to rely more on traditional practices, exhibit lower tolerance, or show less willingness to adopt new strategies. The current results align with findings by [51], who observed a positive association between age and climate change adaptation strategies in Kenya. Similarly, [23] reported that age positively influenced the adoption of some adaptation measures in Ethiopia. However, these findings contrast with those of [52], who found that age decreased the utilization drought-tolerant crop varieties, and [53], who reported that age reduced the uptake of water harvesting techniques as in Kenya.

In terms of gender, being male significantly reduced the odds of not pursuing the specified adaptation options by 97.9%. The substantial decrease in odds suggests that being male decreases the possibility of adopting and implementing adaptation strategies. These findings are consistent with those of [48], who also found that gender decreased the uptake of agronomic practices and livelihood diversification strategies in Ethiopia. However, they contrast with the results of [54], who observed an increase in the uptake of fertilizer and pesticide use with respect to gender in Uganda.

Lack of access to weather information decreased the likelihood of not undertaking the specified adaptation measures. This implies that farmers who have no access to information have a decreased chances adopting available adaptation strategies. Being informed about climate risks and benefits of adaptation likely encourages farmers to take action rather than remain passive. In contrast, the current findings differ from those of [23], who observed a positive link between access to climate information and the adoption of climate change adaptation strategies in Ethiopia. Likewise, [55] also found a positive relationship between access to climate information and the implementation of adaptation measures in the same context.

3.5 Factor Influencing the Number of New Livelihood Activities Selected by Farmers

Table 13 presents the Poisson regression results identifying the factors that influence the number of new livelihood activities adopted by farmers in the study area. The analysis reveals that only two variables—land ownership and group membership—significantly affect the number of new livelihood activities undertaken. Land size owned by the farmer is positively associated with a 3.5% increase the number of new livelihood activities adopted by farmers. Farmers with larger landholdings are generally more inclined to implement new strategies, as they have greater resources and space to test and apply various approaches to address climate change. The current findings accord with the findings of [56], who discovered a positive link between land size and the number of new livelihood activities practiced by farmers in Kenya.

Membership in a farmer group is positively linked to an 11.2% increase in the number of new livelihood activities adopted. Participation in such groups improves farmers' access to vital information, resources, and support networks, which in turn boosts their capacity and willingness to embrace new adaptation strategies. Further, these groups facilitate knowledge exchange, access to new technologies, and encouragement to try new methods. These findings are convergent with those of [57], who established that membership in local organizations boosted the likelihood of adopting climate change adaptation strategies among smallholders farmers in Ethiopia. However, the results contrast with those of [51], who observed group membership reduced the uptake of multiple strategies in Kenya.

4 Conclusion

Farmers in Nakuru County perceive climate changes through various forms, such as higher temperatures, untimely rains, prolonged dry spells, and strong winds. Among these, untimely rains and strong sun were reported as the most common occurrences, while prolonged dry spells and strong sun were seen as having the most significant impact on farmers' livelihoods. Farmers expressed particular vulnerability to hot temperatures and strong sun and anticipate increasing erratic and unpredictable rainfall in the future. Farmers in the study area employed various climate change adaptation strategies, including early planting, livestock keeping, and cultivation of short-season crops, tree planting, irrigation, and participation in off-farm activities. The logit regression analysis revealed that key factors influencing the selection of these strategies included group membership, remittances, access to irrigation, and access to climate-related information, education level, and main income source. Furthermore, results from the Poisson regression model showed that the size of land owned and participation in farmer groups significantly affected the number of new livelihood activities adopted.

Given the important role of group membership in promoting the adoption of adaptation strategies, it is recommended that farmers be encouraged to join or

Table 13 Poisson regression results on factors influencing the number of new livelihood activities selected by farmers

Number of new livelihood activities	Coef.	St. Err.	t-value	p-value	95% Conf	Interval	Sig
Age	0.003	0.002	1.57	0.117	−0.001	0.006	
Gender of respondent	−0.043	0.051	−0.83	0.404	−0.143	0.057	
Gender of household head	0.031	0.058	0.54	0.589	−0.082	0.145	
Education level							
Not finished primary school	−0.062	0.095	−0.65	0.519	−0.249	0.125	
Finished primary school	−0.062	0.097	−0.64	0.52	−0.253	0.128	
Not finished secondary	0.054	0.108	0.50	0.62	−0.158	0.266	
Finished secondary	0.141	0.093	1.51	0.13	−0.042	0.324	
Not finished college/ university	0.2	0.124	1.62	0.106	−0.042	0.443	
Finished college/ university	−0.13	0.124	−1.05	0.295	−0.373	0.113	
Experience							
5 to 10 years	0.09	0.109	0.82	0.41	−0.124	0.305	
> 10 years	0.017	0.107	0.16	0.871	−0.192	0.227	
Household size	−0.008	0.012	−0.69	0.493	−0.031	0.015	
Distance to market	−0.018	0.015	−1.13	0.257	−0.048	0.013	
Income source	0.044	0.027	1.60	0.11	−0.01	0.098	
Remittances	−0.028	0.051	−0.55	0.582	−0.128	0.072	
Land owned	0.035	0.013	2.65	0.008	0.009	0.061	***
Irrigation access	−0.05	0.052	−0.96	0.336	−0.152	0.052	
Group membership	0.112	0.055	2.03	0.042	0.004	0.219	**
Credit access	−0.072	0.068	−1.05	0.293	−0.206	0.062	
Climate information access	−0.076	0.047	−1.60	0.11	−0.169	0.017	
Extension visits							
Private	0.007	0.068	0.10	0.918	−0.126	0.14	
Public	−0.088	0.059	−1.49	0.137	−0.203	0.028	
Public and private	−0.02	0.088	−0.23	0.817	−0.192	0.151	
Constant	0.881	0.175	5.04	0.000	0.538	1.223	***
Mean dependent var		3.038	SD dependent var			1.048	
Pseudo r-squared		0.022	Number of obs			240	
Chi-square		78.925	Prob > chi2			0.000	
Akaike crit. (AIC)		829.295	Bayesian crit. (BIC)			912.830	

*** *p<.01*, ** *p<.05*, and * *p<.1* denotes significance at 1%, 5%, and 10%, respectively
Source: Author's (2025)

establish farmer groups or agricultural cooperatives. These platforms provide opportunities for knowledge sharing, access collective resources, and greater bargaining power to acquire inputs, irrigation equipment, or credit facilities. Additionally, enhancing access to climate information through farmer training, radio programs, mobile apps, or partnerships with local institutions will better equip farmers to make informed decisions about adaptation strategies. Expanding irrigation schemes to increase water availability for irrigation purposes is also important. Lastly, promoting tree planting as an adaptation strategy, alongside other sustainable agricultural practices like agroforestry, should be prioritized.

Acknowledgment The authors sincerely thank the farmers for their cooperation and willingness to provide essential information for this study. Special thanks are also extended to the enumerators who conducted the household surveys, the tour guides, and the officers from the Ministry of Agriculture. Financial support from the European Union's Horizon 2020 research and innovation program and the French ANR was crucial in enabling the successful completion of this work.

Authors' Contribution The first author (LEK) conceptualized the study, developed the research design, created the data collection instruments, conducted field data collection, performed data analysis, and drafted the manuscript. The second author (RCR) contributed to the study design, refined the data collection tools, and coordinated and participated in the data collection process. The third (AS) and fourth (PC) authors contributed by providing input on the study design and reviewing the manuscript with constructive recommendations. The fifth author (DJ) was involved in designing the study, improving the data collection instruments, coordinating and participating in data collection, conducting data analysis, and reviewing and revising the manuscript draft. All authors reviewed and approved the final version of the manuscript.

Data Availability Statement The data that support the findings of this study are available from the corresponding author upon reasonable request. Access to the data will be provided after review and approval by the corresponding author, and data will be shared in accordance with any ethical guidelines or restrictions that may apply.

Conflict of Interest Statement The authors declare that there is no conflict of interest regarding the publication of this paper.

Funding InformationThis project was funded by the European Union's Horizon 2020 research and innovation program under grant agreement No. 862555 (ERA-NET Cofund FOSC) and French ANR under grant agreement ANR-21-FOSC-0003.

References

1. Ceaser, B. J., Omuna, D., Eniru, E. E., Oluwole, A., Mahoro, B., & Musinguzi, D. (2019). Youth's awareness of climate change in Nakuru Town, Kenya. *International Journal of Engineering Science, 3*(11), 19–24.
2. IPCC. (2018). *Global warming of 1.5 °C: An IPCC special report on the impacts of global warming of 1.5 °C above pre-industrial levels and related global greenhouse gas emission pathways*. In Press.
3. FAO. (2016). *The state of food and agriculture. Climate change, agriculture and food security*. FAO.
4. Mugambiwa, S. S., & Tirivangasi, H. M. (2017). Climate change: A threat towards achieving 'sustainable development goal number two' (end hunger, achieve food security and improved nutrition and promote sustainable agriculture) in South Africa. *Jàmbá: Journal of Disaster Risk Studies, 9*(1), 350. https://doi.org/10.4102/jamba.v9i1.350
5. Sloat, L. L., Davis, S. J., Gerber, J. S., Moore, F. C., Ray, D. K., West, P. C., & Mueller, N. D. (2020). Climate adaptation by crop migration. *Nature Communications, 11*, 1243. https://doi.org/10.1038/s41467-020-15076-4
6. Bett, B., Kiunga, P., Gachohi, J., Sindato, C., Mbotha, D., Robinson, T., Lindahl, J., & Grace, D. (2017). Effects of climate change on the occurrence and distribution of livestock diseases. *Preventive Veterinary Medicine, 137*(Pt B), 119–129. https://doi.org/10.1016/j.prevetmed.2016.11.019
7. Shibabaw, A., Alamayehu, G., Desalegn, Y., Adgo, E., Tewodoros, Y., Germer, J., Asch, F., Merene, Y., & Freyer, B. (2014). Exploratory survey on climate change effects, value chain processes, and supportive services: Case study from potato-based farming system of Awi Zone, Ethiopia. *International Journal of Agriculture Innovations and Research, 2*(4) ISSN (Online) 2319-1473.
8. Koskei, E., & Kitetu, J. (2019). Awareness and adaptation to climate change in arid and semi-arid lands: A case of Baringo County. *Proc Envs.*, 24–32.
9. Tesfahunegn, F., Gebreyesus, B., Kirubel, M., & Abadi, T. (2016). Farmer's perception on causes, indicators and determinants of climate change in Northern Ethiopia: implication for developing adaptation strategies. *Applied Geography, 73*, 1–12. https://doi.org/10.1016/j.apgeog.2016.05.009
10. IPCC. (2007). *IPCC 4th assessment report - climate change 2007. Working group II on "impacts, adaptation and vulnerability"*. Available from: http://www.ipcc-wg2.org
11. Atsiaya, G., Ingasia, O., Nakhone, L., & Lagat, J. (2019). Drivers and responses to climate variability by agro-pastoralists in Kenya: The case of Laikipia County. *SN Applied Sciences*, 1. https://doi.org/10.1007/s42452-019-0849-x
12. Taiy, R. J., Onyango, C., & Nkurumwa, A. (2017). Climate change challenges and knowledge gaps in smallholder potato production: The case of Mauche Ward in Nakuru County, Kenya. *International Journal of Agricultural Science and Research, 7*, 719–730. https://doi.org/10.24247/ijasraug201795
13. Codjoe, S., Atidoh, L. K., & Burkett, V. (2011). Gender and occupational perspectives on adaptation to climate extremes in the Afram Plains of Ghana. *Climatic Change*. https://doi.org/10.1007/s10584-011-0237-z
14. Agesa, B. L., Onyango, C. M., Kathumo, V. M., Onwonga, R. N., & Karuku, G. N. (2019). Climate change effects on crop production in Yatta sub-County: Farmer perceptions and adaptation strategies. *African Journal of Food Agriculture Nutrition and Development, 19*(1), 14010–14042.
15. Jiri, O., Mafongoya, P., & Chivenge, P. (2015). Smallholder farmer perceptions on climate change and variability: A predisposition for their subsequent adaptation strategies. *Journal of Earth Science and Climatic Change, 6*, 277. https://doi.org/10.4172/2157-7617.1000277
16. Mwinkom, F. X., Damnyag, L., Abugre, S., & Alhassan, S. I. (2021). Factors influencing climate change adaptation strategies in North-Western Ghana: Evidence of farmers in the

Black Volta Basin in Upper West region. *SN Applied Sciences, 3*, 548. https://doi.org/10.1007/s42452-021-04503-w
17. IPCC. (2021). *Climate change 2021: The physical science basis. Contribution of working group I to the sixth assessment report of the intergovernmental panel on climate change*.
18. Republic of Kenya. (2010). *National Climate Change Response Strategy*. Government Press.
19. Mulwa, R., Rao, K., Gummadi, S., & Kilavi, M. (2015). Effects of climate change on agricultural households' welfare in Kenya. *Climate Research*, 67. https://doi.org/10.3354/cr01357
20. GOK. (2010). *National climate response strategy*. Ministry of Environment and Mineral Resource Development.
21. GoK. (2017). *Kenya climate smart agriculture strategy 2017–2026*. Ministry of Agriculture, Livestock and Fisheries.
22. MoALF. (2016). *Climate risk profile for Nakuru. Kenya County climate risk profile series*. Ministry of Agriculture, Livestock and Fisheries (MoALF).
23. Belay, A., Recha, J. W., Woldeamanuel, T., & Morton, F. J. (2017). Smallholder farmers' adaptation to climate change and determinants of their adaptation decisions in the central Rift Valley of Ethiopia. *Agriculture & Food Security, 6*, 24. https://doi.org/10.1186/s40066-017-0100-1
24. Nyang'au, O. J., Mohamed, H. J., Mango, N., Makate, C., & Wangeci, N. A. (2021). Smallholder farmers' perception of climate change and adoption of climate-smart agriculture practices in Masaba South Sub-county, Kisii, Kenya. *Heliyon, 7*(4), e06789. https://doi.org/10.1016/j.heliyon.2021.e06789
25. Niang, I., Ruppel, O. C., Abdrabo, M. A., Essel, A., Lennard, C., Padgham, J., & Urquhart, P. (2014). Africa. In V. R. Barros (Ed.), *Climate change 2014: Impacts, adaptation, and vulnerability. Part B: Regional aspects*. Cambridge University Press.
26. Abid, M., Schilling, J., Scheffran, J., & Zulfiqar, F. (2016). Climate change vulnerability, adaptation and risk perceptions at farm level in Punjab, Pakistan. *Science of the Total Environment, 547*, 447–460.
27. Kabir, A., Amin, M. N., Roy, K., & Hossain, M. S. (2021). Determinants of climate change adaptation strategies in the coastal zone of Bangladesh: Implications for adaptation to climate change in developing countries. *Mitigation and Adaptation Strategies for Global Change, 26*(7), 30. https://doi.org/10.1007/s11027-021-09968-z
28. Ochieng, J., Kirimi, L., & Mathenge, M. (2016). Effects of climate variability and change on agricultural production: The case of small-scale farmers in Kenya. *NJAS - Wageningen Journal of Life Sciences, 77*(1), 71–78. https://doi.org/10.1016/j.njas.2016.03.005
29. Oremo, O. F. (2013). *Small-scale farmers' perceptions and adaptation measures to climate change in Kitui County, Kenya [Master's thesis]*. University of Nairobi.
30. Dassanayake, W., Mohapatra, S., Luckert, M., & Adamowicz, W. (2017). Households' responses to climate change: Contingent behavior evidence from rural South Africa. *Environment and Development Economics, 23*, 1–26. https://doi.org/10.1017/S1355770X17000328
31. Bateman, I. J., & Turner, R. K. (1993). Valuation of the environment, methods and techniques: The contingent valuation method. In R. K. Turner (Ed.), *Sustainable environmental economics and management: Principles and practice* (pp. 120–191). Belhaven Press.
32. Morton, K. M., Adamowicz, W. L., & Boxall, P. C. (1995). Economic effects of environmental quality change on recreational hunting in northwestern Saskatchewan: A contingent behavior analysis. *Canadian Journal of Forest Research, 25*(6), 912–920.
33. Whitehead, J. C., Phaneuf, D., Dumas, C. F., Herstine, J., Hill, J., & Buerger, B. (2010). Convergent validity of revealed and stated recreation behavior with quality change: A comparison of multiple and single site demands. *Environmental and Resource Economics, 45*(1), 91–112.
34. CIDP. (2018). *Nakuru County integrated development plan 2018–2022*. County Government of Nakuru.
35. Kothari, C. R. (2004). *Research methodology: Methods and techniques* (2nd ed.). New Age International Publishers.

36. Wang, X., & Cheng, Z. (2020). Cross-sectional studies: Strengths, weaknesses, and recommendations. *Chest, 158*(1S), S65–S71. https://doi.org/10.1016/j.chest.2020.03.012
37. Casley, D., & Kumar, K. (1988). *The collection, analysis and use of monitoring and evaluation of data*. The International Bank of Reconstruction and Development.
38. Cochran, W. G. (1977). *Sampling techniques* (3rd ed.). Wiley.
39. Cappellari, L., & Jenkins, S. P. (2003). Multivariate probit using simulated maximum likelihood. *The Stata Journal, 3*(3), 278–294.
40. Greene, W. H. (2008). *Econometric analysis* (6th ed.). Pearson Prentice Hall.
41. Opiyo, F., Wasonga, O. V., Nyangito, M. M., Mureithi, S. M., Obando, J., & Munang, R. (2015). Determinants of perceptions of climate change and adaptation among Turkana pastoralists in northwestern Kenya. *Climate and Development, 8*(2), 179–189. https://doi.org/10.1080/17565529.2015.1034231
42. Asrat, P., & Simane, B. (2018). Farmers' perception of climate change and adaptation strategies in the Dabus watershed, North-West Ethiopia. *Ecological Processes, 7*, 7. https://doi.org/10.1186/s13717-018-0118-8
43. Zeleke, G., Teshome, M., & Ayele, L. (2024). Determinants of smallholder farmers' decisions to use multiple climate-smart agricultural Technologies in North Wello Zone, Northern Ethiopia. *Sustainability, 16*, 4560. https://doi.org/10.3390/su16114560
44. Marie, M., Yirga, F., Haile, M., & Tquabo, F. (2020). Farmers' choices and factors affecting adoption of climate change adaptation strategies: Evidence from northwestern Ethiopia. *Heliyon, 6*, e03867. https://doi.org/10.1016/j.heliyon.2020.e03867
45. Gemeda, O. D., Korecha, D., & Garedew, W. (2023). Determinants of climate change adaptation strategies and existing barriers in Southwestern parts of Ethiopia. *Climate Services, 30*, 100376. https://doi.org/10.1016/j.cliser.2023.100376
46. Kumar, K. N. R., Reddy, M. J. M., Reddy, K. V., Paramesha, V., Balasubramanian, M., Kumar, T. K., Kumar, R. M., & Reddy, D. D. (2023). Determinants of climate change adaptation strategies in South India: Empirical evidence. *Frontiers in Sustainable Food Systems, 7*, 1010527. https://doi.org/10.3389/fsufs.2023.1010527
47. Edward, M. C. (2016). Household socio-economic factors and adoption of climate change adaptation strategies: The case of same district, Tanzania. *"UONGOZI" - Journal of Management Development Dynamics (UJMDD), 27*(1), 85–114.
48. Amare, A., & Simane, B. (2017). Determinants of small-holder farmers' decision to adopt adaptation options to climate change and variability in the Muger Sub basin of the Upper Blue Nile basin of Ethiopia. *Agriculture & Food Security, 6*, 64. https://doi.org/10.1186/s40066-017-0144-2
49. Amir, S., Saqib, Z., Khan, M. I., Ali, A., Khan, M. A., Bokhari, S. A., & Zaman-ul-Haq. (2020). Determinants of farmers' adaptation to climate change in rain-fed agriculture of Pakistan. *Arabian Journal of Geosciences, 13*, 1025. https://doi.org/10.1007/s12517-020-06019-w
50. Sekabira, H., & Qaim, M. (2017). Mobile money, agricultural marketing, and off-farm income in Uganda. *Agricultural Economics, 48*, 597–611. https://doi.org/10.1111/agec.12360
51. Ndiwa, A. M., Mburu, J., Mulwa, R., & Chumo, C. (2024). Determinants of climate change adaptation strategies and intensity of use; micro level evidence from crop farmers in Kenya. *Frontiers in Sustainable Food Systems, 8*, 1376868. https://doi.org/10.3389/fsufs.2024.1376868
52. Kogo, B. K., Kumar, L., Koech, R., & Hasan, M. K. (2022). Response to climate change in a rain-fed crop production system: Insights from maize farmers of western Kenya. *Mitigation and Adaptation Strategies for Global Change, 27*, 50. https://doi.org/10.1007/s11027-022-10023-8
53. Ngetich, F. K., Mairura, F. S., Musafiri, C. M., Kiboi, M. N., & Shisanya, C. A. (2022). Smallholders' coping strategies in response to climate variability in semi-arid agro-ecozones of Upper Eastern Kenya. *Social Sciences & Humanities Open, 6*, 100319.
54. Atube, F., Malinga, G. M., Nyeko, M., Okello, D. M., Alarakol, S. P., Okello-Uma, I., et al. (2021). Determinants of small-holder farmers' adaptation strategies to the effects of climate change: Evidence from northern Uganda. *Agriculture & Food Security, 10*, 6. https://doi.org/10.1186/s40066-020-00279-1

55. Adego, T., & Woldie, A. G. (2021). The complementarity and determinants of adoption of climate change adaptation strategies: Evidence from smallholder farmers in Northwest Ethiopia. *Climate and Development*. https://doi.org/10.1080/17565529.2021.1943296
56. Gebre, G. G., Amekawa, Y., Fikadu, A. A., & Rahut, D. B. (2023). Farmers' use of climate change adaptation strategies and their impacts on food security in Kenya. *Climate Risk Management, 40*, 100495. https://doi.org/10.1016/j.crm.2023.100495
57. Bedeke, S., Vanhove, W., Gezahegn, M., Natarajan, K., & Van Damme, P. (2019). Adoption of climate change adaptation strategies by maize-dependent smallholders in Ethiopia. *NJAS: Wageningen Journal of Life Sciences, 88*, 96–104.

Allelopathic Potential of *Achyranthes aspera* and *Tagetes minuta* Leaf Extracts on Seed Germination of Maize, Millet, Rice, and Sorghum

Jeptoo Loyce, Pascaline Jeruto, Emily Jepkosgei Too, and Lizzy A. Mwamburi

Abstract

Achyranthes aspera is a weed of the family *Amaranthaceae* and grows to about 2 m high. *Tagetes minuta* is weed of the family *Asteraceae* that grows to a height of 1–2 m. The two plants are major weeds of cultivated crops. They have been documented to possess allelopathic activities. Studies have estimated that weeds generate greater losses compared to diseases and pests. That is why this research was carried out to determine the allelopathic potential of *Achyranthes aspera* and *Tagetes minuta* crude leaf extracts on the germination of maize, millet, rice, and sorghum seeds. Leaves were collected from farm fields and shade dried followed by aqueous and ethanolic extraction. Each petri dish contained five seeds that had been surface sterilized. Thirty (30) milliliters of extracts were utilized while distilled water was used as positive control. This was set up at 25 ± 2 °C in three replications adopting a completely randomized design. After 7 days, the lengths of the emerging plumule and radicles were measured, and the germination percentages were calculated. Significant inhibition of radicle and plumule lengths of maize, millet, rice, and sorghum ($p \leq 0.05$) was demonstrated by the extracts. Millet recorded the highest germination percentage (89.44%), while rice recorded the lowest at 67.78%. Generally, the extracts interfered with germination and

Authors Pascaline Jeruto, Emily Jepkosgei Too and Lizzy A. Mwamburi contributed equally to this work.

J. Loyce (✉) · P. Jeruto · E. J. Too · L. A. Mwamburi
Department of Biological Sciences, University of Eldoret, Eldoret, Kenya

N. Ronoh, C. Rono (eds.), *Rethinking Our Ecosystems*, Springer Proceedings in Geography and Sustainability, https://doi.org/10.1007/978-3-031-95781-9_7

growth of these plants. Therefore, when *A. aspera* and *T. minuta* are present in farm fields, there is a substantial probability that maize, millet, rice, and sorghum will grow poorly. This could significantly lower yields, ultimately leading to food insecurity.

Keywords
Allelopathy · Allelochemicals · Weeds · Synergy · Germination

1 Introduction

Achyranthes aspera (devil's horsewhip) and *Tagetes minuta* (wild marigold) are weeds of economic importance distributed throughout tropical and subtropical regions. They can grow to a height of 1–2 m. They can potentially grow on undisturbed waste land and major weed of cultivated crops including maize, sugarcane, and wheat [1]. *Achyranthes aspera* and *Tagetes minuta* have been documented to potentially possess allelopathic activities [1, 2].

A broad category of noneconomic plants, weeds can flourish in areas they are not needed or can coexist alongside crops, causing financial losses such as lower yields as a result of competition and/or the release of allelochemicals, a phenomenon known as allelopathy [3]. A biological phenomena known as allelopathy occurs when one organism produces and releases allelochemicals into the environment, which affect growth, development, germination, and survival of surrounding organisms [4]. Allelochemicals are phytochemicals released from some plants and affect other sensitive plants directly or indirectly. They can have a direct or indirect impact on the growth, development, germination, reproduction, and survival of surrounding plants [5].

Due to competition and/or allelopathy, weeds that grow amid agricultural plants pose a threat to crop production by lowering yields [3]. This is to a higher percentage attributed to allelochemicals. Allelopathic interactions between weeds and crops have been linked to yield reductions in some studies [6]. However, there is little data highlighting the negative impacts of weeds on seed germination due to weed-crop allelopathic interactions, despite estimates that weed-related losses are even higher than those brought about by insect pests and diseases [7]. This is to a higher percentage attributed to the production and release of allelochemicals [2]. These environmentally safe substances, once released into the environment, interact and may influence the growth and development of biological systems, including inhibition and/or stimulation effects [1]. However, their inhibitions effects are often hazardous to the ecosystems.

The aim of this study was to determine the allelopathic potential of *Achyranthes aspera* and *Tagetes minuta* leaf extracts on the germination of maize, millet, rice, and sorghum seeds.

2 Materials and Methods

Plant leaves of *Achyranthes aspera* and *Tagetes minuta* used in this study were identified and then collected from the farm fields of University of Eldoret (UoE). After being brought to the laboratory, these leaves were hand-washed with tap water and then air-dried in a shed until complete dryness was achieved. Following that, the dried leaves were powdered using a grinder to a fine powder, put in sterile containers, labelled, and kept in storage awaiting further analysis.

In separate conical flasks, 30 g of crushed leaf powders from *Achyranthes aspera* and *Tagetes minuta* were soaked for 48 h in 150 ml of ethanol and double-distilled water. Double-distilled water was used to create extract dilutions at concentrations ranging from 10^0 to 10^{-3}. The potential effects of these extracts on seed germination were investigated by placing maize seeds (H6213), millet (pearl millet), rice (NERICA L-19), and sorghum (E-1291) obtained from Kenya seed company in petri dishes. A 1% sodium hypochlorite was used to surface sterilize seeds for 1 min before the germination test, and excess chemical washed off with double-distilled water.

Five (5) seeds which had been surface-sterilized were placed in each petri dish to examine the possible effects of the plant extracts on germination. It was then filled with 30 ml of the test extracts from the experimental plants. To assess possible synergy, 15 ml of each extract was mixed and then dispensed to the petri dishes. Dimethyl sulfoxide-treated petri plates were used as negative controls, and those treated with distilled water were utilized as positive controls. Three replications of each plant extract and test seed type were used in the fully randomized design (CRD) trials. After that, the setup was left in the lab at room temperature (25 ± 2 °C).

After 7 days, the lengths of the radicles and plumules that had developed were measured with a sterile string and recorded on a ruler. After 7 days, the germinated seed numbers was tallied, and the germination percentage was computed using the following formula:

$$\text{Germination percentage}(\%) = \frac{\text{Number of germinated seeds}}{\text{Total number of seeds}} \times 100$$

Kolmogorov-Smirnov test was used to test for normality of data. Analysis of variance (ANOVA) was utilized statistical examination of the effects of plant extracts on plumule and radicle lengths and considered statistically significant at $P \leq 0.05$.

3 Results

3.1 Effects of Ethanolic and Aqueous Extracts on Germination Percentages

After 7 days, the seeds that germinated was counted in order to determine the germination percentages. Millet recorded the highest germination percentage (89.44%),

Table 1 Counts and germination percentage of seeds (source: own contribution)

Plant	*Achyranthes aspera* (A)	*Tagetes minuta* (T)	A + T	Total	Germination%	% of +ve control	% of −ve control
Maize	102	109	104	315	87.50	100	0
Millet	100	107	115	322	89.44	100	0
Rice	75	88	81	244	67.78	98	0
Sorghum	96	109	104	309	85.83	100	0

Table 2 p-values of Kolmogorov-Smirnov test (source: Statgraphics Centurion software)

Plant	Plumule	Radicle
Maize	0.38	0.79
Millet	0.06	0.16
Rice	0.70	0.40
Sorghum	0.62	0.31

while rice recorded the lowest percentage at 67.78%. Maize and sorghum had 87.50% and 85.83%, respectively (Table 1). There was an increase in the seeds germinated with a reduction in extract concentration.

3.2 Test for Normality of Data

Kolmogorov-Smirnov test in Statgraphics Centurion software determined that all the data can be adequately modeled by a normal distribution since all the p-values were greater than or equal to 0.05 (Table 2).

3.3 Effects of Ethanolic and Aqueous Plant Extracts on Plumule Lengths

3.3.1 Maize

Aqueous extracts of *Tagetes minuta* showed the highest percentage reduction in maize plumule lengths (83%), while *Achyranthes aspera* showed the lowest reduction (65%) based on concentration 10^0 (Fig. 1). Ethanolic extracts of *Achyranthes aspera* and *Tagetes minuta* had percentage reductions of 68% and 71%, respectively. A combination of the two extracts resulted in reductions of 71% and 76% for ethanolic and aqueous extracts, respectively.

3.3.2 Millet

The plumule lengths of millet were affected differently with the original concentration (10^0) being more detrimental. The effect was then observed to be reducing with reducing extract concentrations. Ethanolic extracts of *Achyranthes aspera* produced

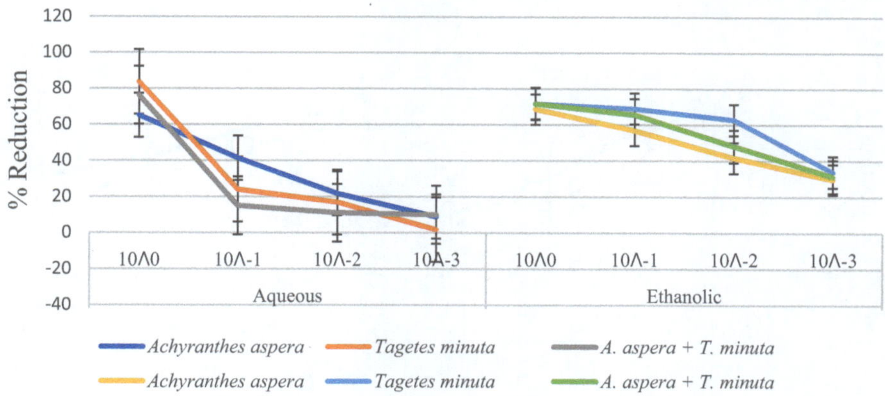

Fig. 1 Percentage reductions in plumule lengths of maize (source: own contribution)

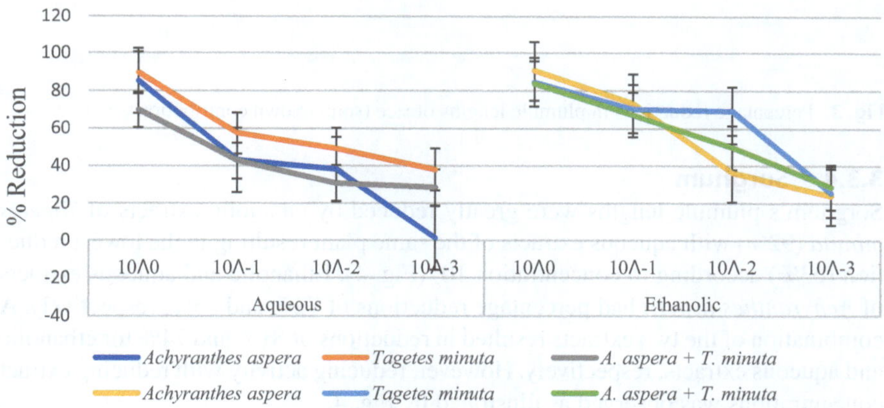

Fig. 2 Percentage reductions in plumule lengths of millet (source: own contribution)

the highest percentage reduction (90%). This study observed lowest percentage reduction from a combination of aqueous extracts from both plants (69%). Ethanolic and aqueous extracts of *Tagetes minuta* had percentage reductions of 84% and 89%, respectively, at concentration 10^0 (Fig. 2).

3.3.3 Rice

The plumule lengths of rice were adversely affected by the extracts of *Achyranthes aspera* and *Tagetes minuta* as well as extracts from a combination of the two plants. This was evident from the percentage reductions of more than 86% in all the tested extracts based on the undiluted concentration (10^0). All the aqueous extracts had percentage reductions of more than 90%. Reduction in the effect of the extracts was noted with reducing extract concentrations, that is, from 10^0 towards 10^{-3} (Figs. 3).

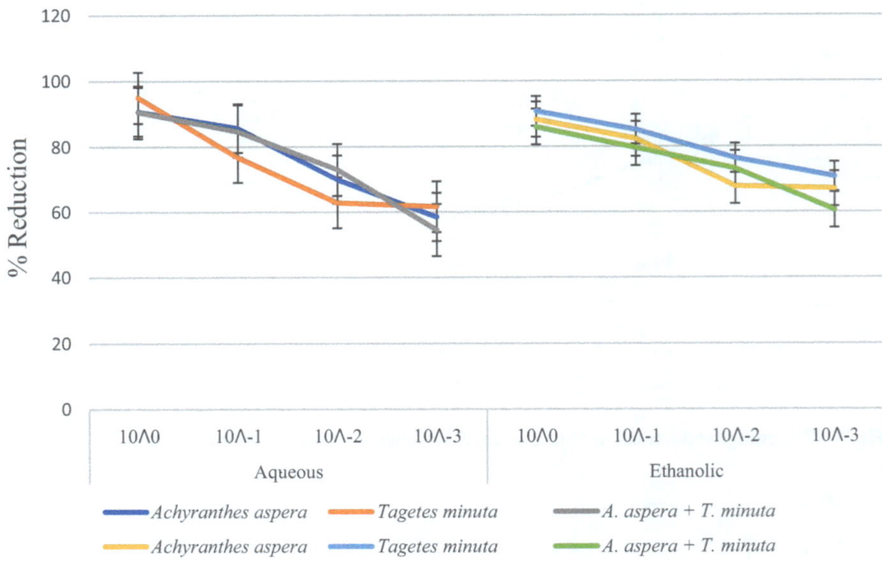

Fig. 3 Percentage reductions in plumule lengths of rice (source: own contribution)

3.3.4 Sorghum

Sorghum's plumule lengths were greatly reduced by ethanolic extracts of *Tagetes minuta* (92%) with aqueous extracts of the same plant resulting in the lowest reduction (63%) according to concentration 10^0 (Fig. 4). Ethanolic and aqueous extracts of *Achyranthes aspera* had percentage reductions of 72% and 83%, respectively. A combination of the two extracts resulted in reductions of 81% and 74% for ethanolic and aqueous extracts, respectively. However, reducing activity with reducing extract concentrations was observed as illustrated by Fig. 4.

ANOVA determined that the plumule lengths of the test plants were statistically significantly impacted by the main factors (extraction concentration, solvent used for extraction, plant extract utilized, and each plant species examined) and their interactions at $P < 0.05$ (Table 3).

Given all the p-values were less than 0.05, ranging from 0.0000 to 0.0004 as at $P \leq 0.05$ (Table 3), these factors had statistically significant effects on plumule length of maize, millet, rice, and sorghum. This suggests these plant extracts impacts the germination of the tested plants and therefore should be controlled given they grow as weeds.

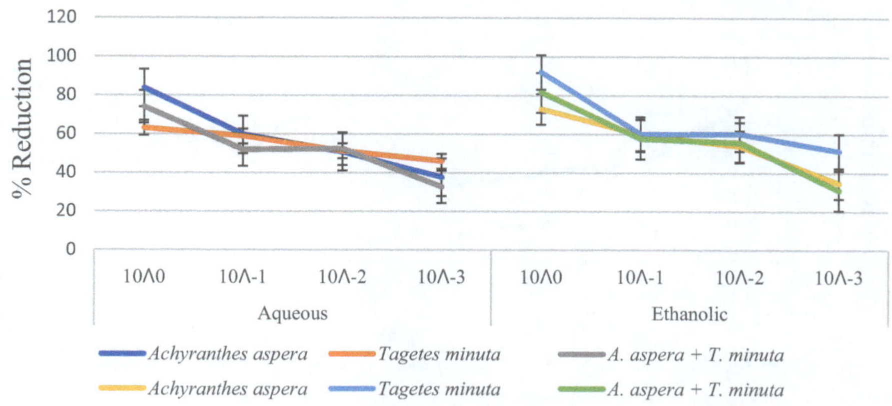

Fig. 4 Percentage reductions in plumule lengths of sorghum (source: own contribution)

Table 3 Effects of extracts on plumule lengths (source: Statgraphics Centurion software)

Source of variation	F-ratio	P-value	Effect as at $p \leq 0.05$
Concentration	287.04	0.0000	Significant
Solvent used	12.46	0.0004	Significant
Plant extract used	13.32	0.0000	Significant
Plant species	222.16	0.0000	Significant
A: *Achyranthes aspera*	29.2	0.0000	Significant
T: *Tagetes minuta*	39.7	0.000	Significant
A + T	46.8	0.0000	Significant

3.4 Effects of Ethanolic and Aqueous Plant Extracts on Radicle Lengths

3.4.1 Maize

The highest percentage reduction in the lengths of maize seeds radicles was recorded on aqueous extracts of *Tagetes minuta* (83%) in the original concentration. Ethanolic extracts of both *Achyranthes aspera* and *Tagetes minuta* as well as aqueous extracts of a combination of the two plants had percentage reductions of 78% each. Aqueous extracts of *Achyranthes aspera* had a percentage reduction of 79%, while ethanolic extracts of a combination of the two plants investigated resulted in a reduction of 80% (Fig. 5).

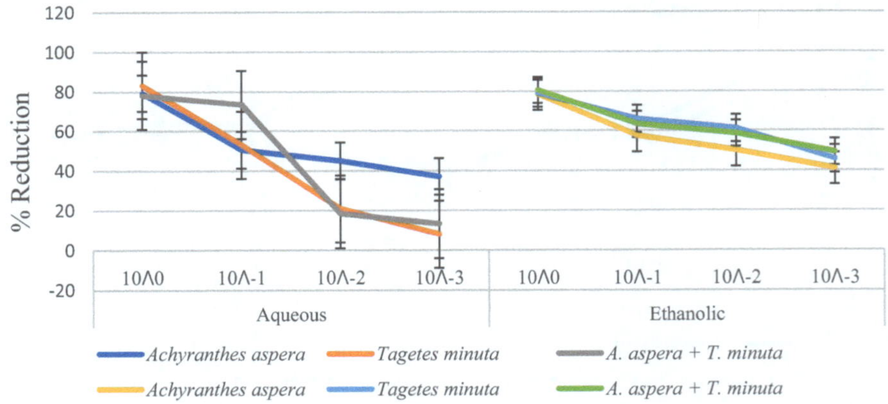

Fig. 5 Percentage reductions in radicle lengths of maize (source: own contribution)

3.4.2 Millet

The radicle lengths of millet were more affected by the extracts when compared to its plumules. This was evident from the percentage reduction of 94% at the highest and 79% at the lowest based on concentration 10^0. Those were obtained from aqueous extracts of *Achyranthes aspera* and ethanolic extracts of a combination of the two plants, respectively. Reductions in the lengths of radicle of 82% and 91% by the effect of *Tagetes minuta* aqueous and ethanolic extracts was noted in millet seeds (Fig. 6). However, the effect of *Achyranthes aspera*'s ethanolic extracts as well as aqueous extracts from a combination of the two plants was similar and could only reduce the lengths by 84% and 85%, respectively (Fig. 6).

3.4.3 Rice

Both aqueous and ethanolic extracts from a combination of the two plants under study showed the lowest percentage decrease in the lengths of rice seed radicles (77%), while aqueous extracts of *Tagetes minuta* showed the largest percentage reduction (96%) at concentration 10^0. Aqueous and ethanolic extracts of *Achyranthes aspera* yielded percentage reductions of 80% and 82%, respectively, while ethanolic extracts of Tagetes *minuta* could only reduce the radicle lengths of rice by 79% (Fig. 7).

3.4.4 Sorghum

According to concentration 10^0, the highest percentage reduction in the lengths of sorghum seeds radicles was obtained from ethanolic extracts of *Tagetes minuta* (97%) with the lowest being from both aqueous and ethanolic extracts of the combination of the two plants investigated (87%). Aqueous and ethanolic extracts of *Achyranthes aspera* yielded percentage reductions of 96% and 87%, respectively, while aqueous extracts of *Tagetes minuta* could only reduce the radicle lengths of rice by 80% (Fig. 8).

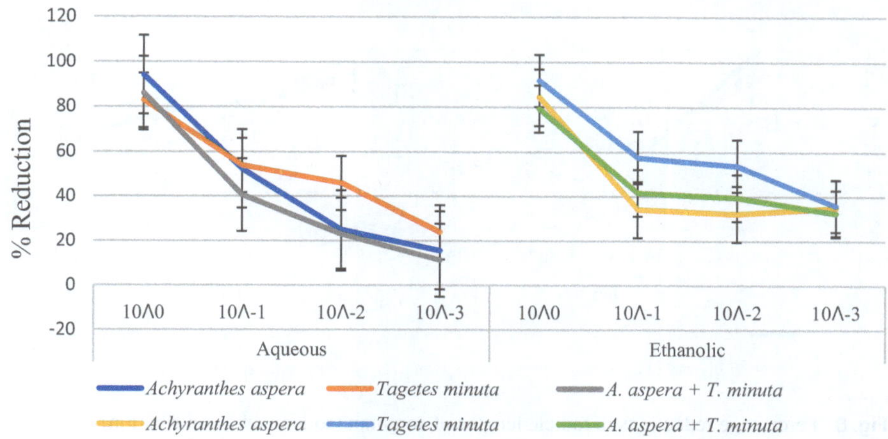

Fig. 6 Percentage reductions in radicle lengths of millet (source: own contribution)

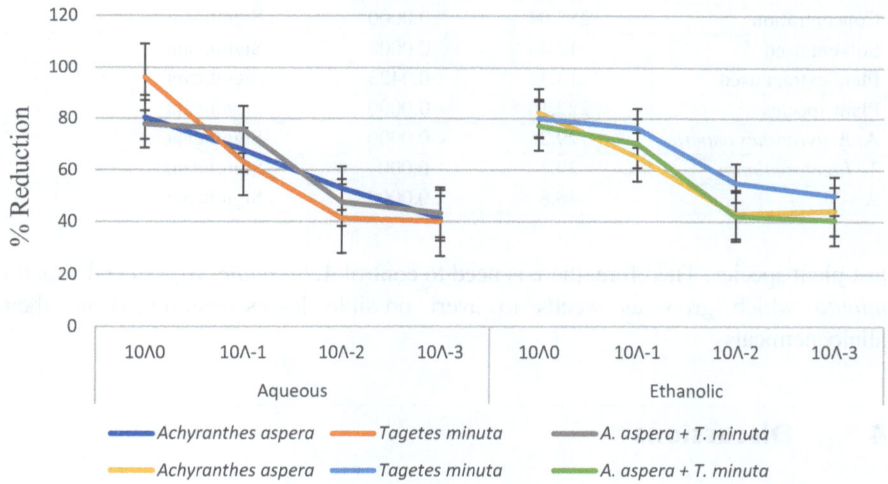

Fig. 7 Percentage reductions in radicle lengths of rice (source: own contribution)

ANOVA was used to determine the effects of ethanolic as well as aqueous plant extracts on the radicle lengths of maize, millet, rice, and sorghum. Since all p-values are less than 0.05, ranging from 0.0000 to 0.0423 at 95.0% confidence level ($P \leq 0.05$), it then implies that these factors had statistically significant effects on radicle length at ($P \leq 0.05$) (Table 4).

The main factors (concentration of the extract, solvent used, plant extract used, and the plant type under investigation) and the individual plant extracts (*Achyranthes aspera* and *Tagetes minuta*) and their interaction had a statistically significant effect on radicle length (Table 4). This therefore means that the extracts of the plants under investigation can significantly inhibited germination and possibly the growth of the

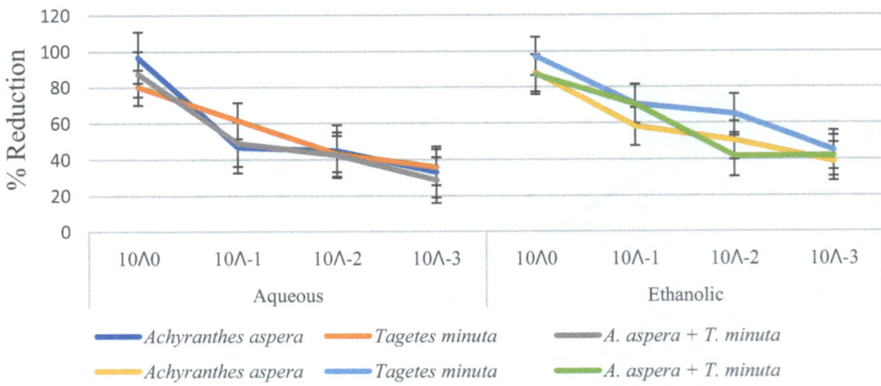

Fig. 8 Percentage reductions in radicle lengths of sorghum (source: own contribution)

Table 4 Effects of extracts on radicle lengths (source: Statgraphics Centurion software)

Source of variation	F-ratio	P-value	Effect as at $p \leq 0.05$
Concentration	287.04	0.0000	Significant
Solvent used	12.46	0.0000	Significant
Plant extract used	13.32	0.0423	Significant
Plant species	222.16	0.0000	Significant
A: *Achyranthes aspera*	29.2	0.0000	Significant
T: *Tagetes minuta*	39.7	0.000	Significant
A + T	46.8	0.0000	Significant

test plant species. Therefore, there is need to control *Achyranthes aspera* and *Tagetes minuta* which grow as weeds to avert possible losses resulting from their allelochemicals.

4 Discussions

Allelopathic activities of aqueous and ethanolic crude leaf extracts of *Achyranthes aspera* and *Tagetes minuta* was tested on maize, millet, rice, and sorghum seeds. These are commonly grown field crops which are members of *Gramineae* family. The findings of this study showed that *Achyranthes aspera* and *Tagetes minuta* crude leaf extracts reduced the germination percentages, radicle length, and the plumule lengths of the test cereal crop seeds. This is corroborated by the data of plumule as well as the radicle lengths which was significantly reduced when compared with control. Statistical analysis using ANOVA returned all the p-values being less than 0.05 at 95.0% confidence level ($P \leq 0.05$). That suggests that the radicle and plumule lengths were clearly affected significantly. It was also evident that there was reducing activity with reducing extract concentration which was also noted by [6, 8].

According to earlier research, weed extracts may contain phototoxic compounds and plant growth inhibitors that contribute to allelopathy [8]. They can be growth regulators, some nutrients, phytochemicals, and toxins [6, 9]. They may also contain various phytotoxic compounds that potentially work as herbicides to inhibit germination and growth of other plant species [1]. Their role in allelopathy is however not been well characterized to date.

In this study, ethanolic plant extracts had a greater allelopathic activity when compared to aqueous based extracts. The radicle lengths also appeared to be affected more compared to the plumule lengths. This observation was also made by [6]. These findings can be attributed to the allelopathic potential from extracts of *Achyranthes aspera* and *Tagetes minuta*. Similar information has also been documented by other authors [1, 2, 10, 11] among others.

Elsewhere, extract of *A. aspera* was demonstrated to be successfully in inhibiting the emergence of pearl millet, guar, and maize [10]. [12] also reported the effect of *Parthenium hysterophorus* L. in maize. In petri plates, *A. aspera* was found to inhibit germination percentage, root length, and dried biomass of rice [11]. [13] also documented the allelopathic influence *Achyranthes aspera* (L.) on the germination and growth behavior of *Oryza sativa* (L.). [14] reported a significant decrease in the germination index of pearl millet, sorghum, and maize in response to aqueous extracts of various plant parts of *A. aspera*.

T. minuta has also been shown to develop a weedy habit and compete with several commercially significant crops, such as beans, maize, and rice [15]. In petri plates, the extracts of *Andrographis paniculata*, *Plectranthus amboinicus*, and *Tagetes minuta* adversely affected the shoot and root length of rice, cowpea, and green gram [16].

The effect of *A. aspera* and *T. minuta* among other weeds on germination, growth, and yields of major crops like wheat, sorghum, pearl millet, maize, and rice among others has been documented by [2, 10–12, 15–18], among others.

The results of this study have revealed that either *Achyranthes aspera* and *Tagetes minuta* either singly or in combination exhibit allelopathic properties on germination behavior of the tested crops and can consequently affect plant stand and production. Therefore, when *A. aspera* and *T. minuta* are present in farm fields, there is a substantial probability that maize, millet, rice, and sorghum will grow poorly. This could significantly lower yields, cause farmers to lose money, and ultimately lead to food insecurity.

5 Conclusions and Recommendations

Our research has shed light on the allelopathic effects of ethanol as well as aqueous extracts from *Achyranthes aspera* and *Tagetes minuta* on seed germination of maize, millet, rice, and sorghum. Ethanolic-based extracts conferred a relatively higher activity when compared to aqueous-based extracts. The extracts have also showed the effects of weeds on the germination of seeds from the plants under investigation which is likely to impact growth and yields of crops in the farm field where they grow. The plant extracts were more detrimental on rice followed by maize and then sorghum with minimal effect on millet.

This study recommends preemergence control of the weeds in order to achieve high germination, good plant establishment, and ultimate good yields of the cereal crops tested. Since this study only explored seedling germination behavior, the potential effects of the extracts on the growth, development, and production can be explored. Additional research is also required to identify the weeds' inhibitory allelochemicals, which may be utilized to create eco-friendly products that will prevent the growth of other plant species.Data AvailabilityThe study was part of a thesis submitted to the University of Eldoret. Data will soon be available at the university repository but can also be availed in excel sheets anytime by the corresponding author if they are required.

References

1. Tanveer, A., Safdar, M., Farooq, N., Sudozai, M., Nadeem, M., & Abbas, T. (2019). Exploring the herbicidal potential of *Achyranthes aspera* against some weeds. *Planta Daninha, 37*, e019188205.
2. Arora, K., Batish, D. R., Singh, H. P., & Kohli, R. K. (2015). Allelopathic potential of the essential oil of wild marigold (*Tagetes minuta* L.) against some invasive weeds. *Journal of Agriculture and Environmental Sciences, 3*, 56–60.
3. Anwar, T., Ilyas, N., Qureshi, R., & Malik, M. A. (2019). Allelopathic potential of *Carica papaya* against selected weeds of wheat crop. *Pakistan Journal of Botany, 51*(1) Available from: http://pakbs.org/pjbot/paper_details.php?id=8279
4. Bahadur, S., Verma, S., Prasad, S., Madane, A., Maurya, S., Gaurav, V., et al. (2015). *Eco-friendly weed management for sustainable crop production—A review*.
5. Nazir, T., Uniyal, A., & Ahmed, M. (2014). Allelopathic response of medicinal plants on germination and growth of traditional field crops. *Indian Forester, 140*, 695–700.
6. Kemboi, M. K., Jeruto, P., Mwamburi, L., & Korir, R. (2022). Allelopathic potential of *Centella asiatica* leaves on seed germination and seedling growth of selected field crops. *Africa Environmental Review Journal, 5*, 2.
7. Zohaib, A., Abbas, T., & Tabassum, T. (2016). Weeds cause losses in field crops through allelopathy. *Notulae Scientia Biologicae, 8*(1), 47–56.
8. Raoof, K. A., & Siddiqui, M. B. (2012). Evaluation of allelopathic impact of aqueous extract of root and aerial root of *Tinospora cordifolia* (Willd.) Miers on some weed plants. *Analele Universităţii din Oradea - Fascicula Biologie, 1*, 29–34.
9. Alagesaboopathi, C. (2018). Allelopathic potential of aqueous extracts of *Centella asiatica* (L.) urban on germination and seedling growth of Oryza sativa L. Varieties from Tamilnadu, South India. *International Journal of Pharmacy and Biological Sciences*, 7.
10. Safdar, M. E., Aslam, A., Qamar, R., Ali, A., Javaid, M. M., Hayyat, M. S., et al. (2021). *Allelopathic effect of prickly chaff flower (Achyranthes aspera L.) used as a tool for managing noxious weeds*.
11. Sidhu, H. K., Malik, I. A., Varun, M., D'souza, R. J. (2021). *Allelopathic effect of leaf extracts of some common weeds on seed germination characteristics and growth of Oryza sativa L*, 475–81.
12. Safdar, M. E. (2015). *Interference and management studies of Parthenium hysterophorus L. in maize*.
13. Sharma, M., & Satsangi, G. (2012). Allelopathic influence of aqueous extract of some selected weeds on seed germination and growth of cultivated species (*Oryza sativa* L.). *Indian Research Journal Of Genetics and Biotechnology, 4*(03), 163–166.
14. Tanveer, A., Safdar, M. E., Tariq, M. A., Yasin, M., & Noorka, I. R. (2014). Allelopathic inhibition of germination and seedling vigor of some selected crops by *Achyranthes aspera* L. *Herbologia, 14*(2).
15. Arora, K., Batish, D., Kohli, R., & Singh, H. (2017). Allelopathic impact of essential oil of *Tagetes minuta* on common agricultural and wasteland weeds. *Innovare Journal of Agricultural Sciences, 5*, 1–4.
16. Shakkira, K. (2022). *Allelopathy for weed management in field crops*.
17. Sadia, S., Qureshi, R., Khalid, S., Nayyar, B. G., & Zhang, J. (2015). Role of secondary metabolites of wild marigold in suppression of Johnson grass and Sun spurge. *Asian Pacific Journal of Tropical Biomedicine, 5*(9), 733–737.
18. Patil, H., & Sharma, R. (2021). Allelopathic potential of *croton bonplandianus* bail. *Journal of Pharmacy & Bioallied Sciences, 9*(1), 29.

Green ICT's Emerging Trends Toward Sustainability in Mitigation and Adaptation to Climate Change in Great Rift Valley

Irine C. Samoei, Fancy Kiptoo, Lucy Kemboi, and Godfrey Kipngetich

Abstract

Climate change is a critical global issue, driven by rising greenhouse gas (GHG) emissions from human activities such as fossil fuel consumption, deforestation, and industrial processes. The consequences are extreme weather, rising sea levels, and biodiversity loss which threatened ecosystem and human societies. The concept of green ICT or green computing plays a key role in both mitigation (reducing GHG emission) and adapting (building people's resilience to climate change effects). Through the implementation of green computing, organizations become more sustainable by reducing energy consumption, minimizing electronic waste, and optimizing resource utilization. Therefore, it is important for individuals, businesses, and governments to promote and invest in green computing practices and technologies to ensure a sustainable future. This paper assesses green ICT emerging trends, with specific focus on circular economy initiatives, green development plans, and the implementation of green ICT trends in selected countries to enable them to monitor, mitigate, and adapt to climate changes within the Great Rift Valley countries. This study is achieved through a paper review in green ICT and circular economy areas. The future of green computing focuses on reducing energy consumption and carbon emissions, increasing the

Authors Fancy Kiptoo, Lucy Kemboi and Godfrey Kipngetich contributed equally to this work.

I. C. Samoei (✉) · F. Kiptoo · G. Kipngetich
Department of Information and Communication Technology, Moi University, Eldoret, Kenya
e-mail: irenechepkoech@mu.ac.ke; fancyjep@mu.ac.ke; godfrey_ronoh@mu.ac.ke

L. Kemboi
Institute of Open and Distance Learning, Moi University, Eldoret, Kenya
e-mail: lucykemboi@mu.ac.ke

N. Ronoh, C. Rono (eds.), *Rethinking Our Ecosystems*, Springer Proceedings in Geography and Sustainability, https://doi.org/10.1007/978-3-031-95781-9_8

use of renewable energy sources, and promoting the circular economy. As technology continues to evolve, the findings will be of great importance in guiding the design and operations of technology in a way that minimizes its environmental impact while combating climate change.

Keywords
Climate change · Greenhouse gas · Mitigation · Adaptation · Green ICT · Resilience

1 Introduction

Disruptions caused by climate change have affected the planet; this variance has been a reflection of natural events for decades. The impacts vary by location, with high latitudes seeing more global warming than tropical regions. While some parts of the Great Rift Valley have extended dry seasons or little rain, others have significant rains. Mitigating the impact of climate change must be of great importance, putting into consideration that this global threat requires international commitment to address its dire consequences to ensure global sustenance. Green ICT is a key enabler of climate mitigation and the transition to economic systems that aims for sustainable development while reducing economic risk and ecological damage.

1.1 Roles of Green ICT in Promoting Climate Change [4]

1.1.1 Actions on Adaptation to Climate Change

Proactive steps must be taken at the local and national levels to manage the effects of climate change. ICT is essential to this process because it makes it possible for early warning and disaster monitoring systems, improved communication for disaster management, environmental education and awareness, and green ICT solutions. In the face of climate change, communities may increase resilience, save lives, and maintain vital resources by increasing access to contemporary ICT-driven solutions, especially in developing nations.

1.1.2 Actions to Reduce GHG Emissions

The United Nations Framework Convention on Climate Change (UNFCCC) states that the ICT sector emits emissions at every stage of its life cycle, but especially during the manufacturing, raw material extraction, and embodiment phases. Research has shown that adoption of ICT services and circular economy principles is essential in mitigating the effect in other sectors. By maximizing energy efficiency, putting automatic power management into place, giving priority to energy-efficient hardware, and prolonging device lifespans, smarter ICT can assist lower carbon footprints. By adopting these measures, organizations can significantly minimize ICT-related environmental impacts while maintaining technological progress.

1.1.3 Actions on Mitigation of Climate Change

Climate change mitigation focuses on reducing greenhouse gas (GHG) emissions by cutting their sources or enhancing carbon sinks. The adoption of green ICT plays a key role in various sectors, as highlighted in the **Global e-Sustainability Initiative (GeSI) Smart 2020 report** examples include smart technologies used in transforming industries by enhancing efficiency, reducing energy consumption, and cutting carbon emission of up to 30%. By integrating ICT into various industries, we can enhance energy efficiency, cut carbon footprints, and drive sustainable development while maintaining productivity and connectivity. IT infrastructure powered by renewable energy is becoming more and more popular as businesses realize the financial and environmental advantages. Data centers and IT systems are increasingly powered by wind and solar energy, which lessens dependency on fossil fuels and carbon emissions. As renewable energy adoption expands, IT infrastructure will become more sustainable, cost-effective, and environmentally friendly, helping mitigate climate change while ensuring reliable digital operations. Based on East Africa Internet Governance Forum (IGF) presentation, it is highlighted that the emerging ICT solutions are increasingly leveraging solar-powered routers and green technology to promote digital inclusion while tackling climate change [13]. ICT solutions play a vital role in energy sustainability by improving efficiency, optimizing resource use, and enabling green technology adoption. These solutions help both the ICT sector itself and other carbon-intensive industries transition to low-carbon models such as green ICT for energy sustainability, ICT-Driven green technology, and digitization for sustainability [6]. This study explored the mapping of green ICT's to adaptation, monitoring, and mitigation of climate changes in the Great Rift Valley with focus on analyzing the circular economy initiatives in selected Greater Rift Valley countries and assessing the green development action plan and the implementation of green ICT emerging trends.

2 Literature Review

2.1 Green ICT

Green ICT refers to a clear process that focuses on making the ICT sector environmentally friendly as well as promoting sustainability through ICT. This includes utilizing ICT to reshape and improve various socioeconomic sectors [32]. The goal of green ICT is to address environmental issues by adopting sustainable practices throughout the entire life cycle of ICT products and services [17]. The ICT sector has a major environmental impact contributing to carbon emissions and electronic waste (e-waste).

Green technologies have the potential to drive economic transformation in Africa by tackling youth unemployment, promoting sustainable development, and accelerating economic growth [11]. In East Africa, green ICT's emerging technologies are shaping a sustainable prospective future by optimizing resources, cutting costs, and enabling digital transformation. Kenya, Rwanda, and Tanzania have embraced emerging technologies like artificial intelligence (AI), blockchain, IoT, biotechnology, and renewable energy to drive innovation and solve societal challenges across various sectors as indicated in Table 1 [12].

2.2 Circular Economy

The circular economy is all about minimizing waste and maximizing use of resources by designing sustainable products that allow reuse and recycling as long as possible [7]. In green ICT context, the circular economy emphasizes designing electronic products and digital infrastructure that can be easily disassembled, repaired, upgraded, and recycled at the end of their useful life [14]. By merging circular economy principles with digital transformation, we can create a more sustainable and efficient approach to managing electronic waste(e-waste) while fostering innovation, resources, efficiency, and job creation [5]. The table below discusses, in particular, the most prominent ICT-enabled technologies and how those technologies influence CE.

Over the last 5 years, mobile operators in Africa have invested nearly $45 billion in capital expenditures (CapEx), with the majority of this investment going into 4G network expansion. This significant investment reflects the growing demand for faster, more reliable mobile connectivity across the continent. With the introduction of 5G fixed wireless access (FWA), the Internet speed has increased more than 10x compared to 4G FWA [30, 31]. 5G FWA allows resources to be shared between mobile broadband, and FWA service examples in East Africa includes Safaricom and Vodacom. Penetration as of year 2022 is as follows: Kenya 1.4%, Tanzania 0.1%, and Rwanda 0.1% [31]. Based on the 2013 study done by Annual Europe, Middle East, and Africa (EMEA) WAN managers being a key industry report analyzing wide area network (WAN) trends, challenges, and strategies across businesses in the EMEA region, Kenya emerged as a leading early adopters of M2M communication technologies in industries that primarily required real-time monitoring, automation, and remote management in Africa. This was driven by a strong digital infrastructure, mobile penetration, and fintech innovation. This trend shows a greater focus on automation, IoT, and data-driven decision-making in Kenya's enterprise sector. Great Rift Valley countries, such as Kenya, Tanzania, Uganda, Rwanda, and Ethiopia, are increasingly adopting solar-powered routers to enhance internet connectivity while promoting green ICT solutions. These routers play a crucial role in reducing the carbon footprint of Internet infrastructure and bridging the digital divide in rural and off-grid areas. By leveraging renewable energy, these routers enable sustainable connectivity, supporting education, healthcare, e-commerce, and digital inclusion [13]. The development of solar-powered routers aligns with the global push for green innovations that are accessible to rural and low-income communities. By combining renewable energy and internet connectivity, this technology helps bridge the digital divide, ensuring that no one is left behind in the digital age. Cloud computing, edge computing, artificial intelligence, and energy

Table 1 Green ICT emerging technologies use cases

Technologies	Sectors	Kenya [12, 20, 30]	Rwanda [21, 22]	Tanzania [18, 27]
Artificial intelligence (AI)	Healthcare	Ilara health, Toto health and SIHA AI	Babyl and Zipline	Dawa Mkononi
	Agriculture	Sun culture, UjuziKilimo, and PlantVillage Nuru	Precision agriculture, smart Nkunganire, and smart irrigation	Mazao hub
	Finance	Tala, branch, and Safaricom MPESA	Africa development Bank, Bank of Kigali, and central Bank of Rwanda	Mpesa, Tigo Pesa, Airtel money, Azam Pesa
	Education	M-Shule and Eneza education	ACEIoT student mobile apps and student attendance monitoring systems	Silabu, smart Darasa, Shule Yetu, Shule direct
Internet of things (IoT)	Agriculture	SunCulture	e-Soko and smart potato	65 CISCO solar-powered IoT sensors have been installed at the Tunasikia farm in Utengule, Tanzania; Mobile Kilimo
	Smart home	Schneider electric, Davis & Shirtlift, and Grandfov company	Smart gas and smoke detectors, smart PIR motor sensors, and smart lighting switches and outlets	Bagamoyo Smart City
	Energy management	M-KOPA solar and Kenya power smart grid, smart meter, microgrid, and PayG cooks	BaseN platform for remote energy monitoring, SLS energy's battery-as-a-service. NODE red IoT-based intelligent energy efficiency Management Systems for Smart Industries (IoT-IEEMS), and ACEIoT IoT-based net metering monitoring system	IoT solutions by Vodacom Tanzania, Devergy's solar energy monitoring in rural areas
	Finance and banking	Equity Bank and KCB Bank	Bank of Kigali	Manka, bank of Tanzania

(continued)

Table 1 (continued)

Technologies	Sectors	Kenya [12, 20, 30]	Rwanda [21, 22]	Tanzania [18, 27]
	Transport	Little, cabs, Fara s, Wasili, and Indrive	Tap & go and Innovision ltd. (Yego moto and YegoCabs)	Infrastructure monitor
Big data	Healthcare	Patient record analysis and PharmAccess	Rwanda integrated health information system, GSM module, digital-first integrated care, and ANURA	Healthcare delivery
	Transport and logistics	Glovo Kenya and Maersk line	Tap & go, safe motors, and Yego innovision ltd. (Yego moto and YegoCabs)	Tanzania National Roads Agency (TANROADS), Tanzania railways corporation (TRC) and port of Dar Es Salaam
	Ecommerce and retail	Jumia and Kilimall	Rwandair, REG S Ltd, Kasha, Jumia, Yubeyi.com, Pikko Stores, Ho Meness Rwanda, Ihaha Technologies, HeHe Limited, Different Hotels and Transport Companies	Jumia, Shop Online Tanzania, KiKUU and Zudua, Piki, Paisha, Kupatana, Inalipa, and duka. direct, FUNDI
	Government service delivery	E-citizen	UMUCYO: Rwanda on-line E-procurement system Irembo: E-government services platform Rwanda electronic cargo tracking system (RECTS)	Government Mobile platform citizens portal (m-GOV), government e- payment gateway (GePG), government communication network (GovNeT) and government mailing system (GMS), Parliament online information system (POLIS), e-Ga help desk, Posta Kiganjami
Blockchain technology	Fintech sector	Crypto currency and peer-to-peer trading platform	Kotani pay	Tunzaa, Mipango, Bizzyn

(continued)

Table 1 (continued)

Technologies	Sectors	Kenya [12, 20, 30]	Rwanda [21, 22]	Tanzania [18, 27]
	Humanitarian sector	Digital identification creation, transaction tracking, and data integrity	IECMS justice system	Land ownership, secure digital identities, and cross-border payment, e-Vibali
	Agriculture and supply chain	Agunity, Ngeni labs, and Wymbee	INATrace for coffee supply chains, farmer Connect's blockchain solution, Circulor's mineral traceability	Tanzania breweries limited (TBL) and BanQu partnership, blockchain for fertilizer subsidy distribution
	Healthcare sector	MYDAWA and PharmAccess foundation	Geriatrics, management of chronic diseases, biomedical and pharmaceutical industries for research and clinical practice	Antenatal care monitoring with blockchain

Source: [12, 18, 22, 27], and author's own contribution

efficient computing are among the latest green ICT emerging trends that have circular economy elements utilized in selected countries as indicated in Table 1.

2.3 Green Development Action Plan

2.3.1 The Role of Policy and Regulation

Government policies and regulations play a crucial role in promoting green ICT. Several countries and regions have implemented policies to encourage the adoption of sustainable practices in the ICT sector. Green development action plan: This plan outlines goals for reducing the environmental impact of ICT products, including targets for energy efficiency, e-waste recycling, and the use of sustainable materials [23].

Policy Development on Circular Economy in Africa [10]

1. National CE policies align with broader sustainable development strategies and support long-term environmental and economic resilience.
2. Product policies that focus on reducing waste, encouraging recycling, and minimizing environmental impact, particularly for high-pollution materials like plastics, electronics, and packaging.

3. Extended producer responsibility (EPR) is a policy framework that places the responsibility for the environmental impact of products throughout their entire lifecycle on producers.
4. Waste management and recycling policies reduce waste, encourage resource recovery, and minimize environmental impact by addressing every stage of the waste life-cycle, from generation to disposal.
5. Fiscal policies play a crucial role in promoting the circular economy (CE) by using taxation, subsidies, and government spending to encourage sustainable practices. These policies help reduce waste, promote recycling, and drive innovation in circular business models.

2.3.2 Circular Economy Policies in Great Rift Valley

In the Greater Rift Valley, many African governments have already factored CE into their national development planning tools. Waste management and recycling policy, Extended Producer Responsibility policy (Uganda, Rwanda, Tanzania, and Mozambique), and National Circular Economy Policy in Kenya and Burundi [15]. Rwanda has emerged as a leader in advancing the **circular economy**, particularly through its **Green Growth and Climate Resilience Strategy** (**GGCRS**) [29]. This strategy focuses on sustainable development by promoting the following:

Energy and water efficiency: encouraging industries to adopt resource-efficient technologies

Green infrastructure: implementing eco-friendly site preparation, building designs, and construction methods

Sustainable waste management: developing waste treatment and recycling solutions within special economic zones (SEZs) and industrial parks

By integrating circular economy principles, Rwanda is setting an example for sustainable industrialization, reducing environmental impact while fostering economic growth [19].

A review of existing literature indicates that Tanzania currently lacks a specific policy or regulation dedicated to e-waste management. However, several broader environmental and industrial policies touch on aspects of waste management and sustainability. As more organizations prioritize sustainability, we can expect to see more innovations in green computing in the years to come.

2.3.3 Regulations and Policy for Waste Electrical and Electronic Equipment (WEEE)

The Global E-waste Monitor 2020 highlights the rapid growth of electronic waste (e-waste) worldwide. The report states that the world generated 53.6 million metric tons (Mt) of electronic waste (e-waste) in 2019, and if current trends persist, e-waste is expected to reach 74.7 million metric tons by 2030. The rise in electronic waste (e-waste) is a growing environmental crisis driven by higher consumption of electronics, shorter product lifespans, and limited recycling efforts [8]. Africa is the consumer of ICT products and has also implemented policies to promote green

ICT. Several circular economy initiatives are currently underway in Africa. Africa e-waste was 2.9 metric tons in 2021. Based on figures highlighted in East Africa community organization report, the growing e-waste challenge in East Africa shows how electronic waste generation is increasing over time. The figure are as follows: Tanzania (2017): 4750 metric tons (according to Tanzania National Bureau of Statistics), Kenya (2021): 5100 metric tons, and Rwanda (2021): 1500 metric tons [24]. Kenya's Sustainable Management Act 2022 is in place, and it provides a clear guidelines, responsibilities, and enforcement mechanisms to ensure the safe handling, collection, transportation, recycling, and disposal of e-waste. With the growing e-waste challenge in Rwanda, the 2018 national e-waste policy has set a foundation for sustainable e-waste management, driving economic opportunities, digital inclusion, and environmental protection [1]. Rwanda has implemented a comprehensive approach to electronic waste (e-waste) management, refurbishment, and digital access so as to demonstrate how ICT and circular economy principles can drive sustainability, economic development, and social inclusion [30]. An insight from an environmental engineer at National Environment Management Council (NEMC), Tanzania, highlighted key challenges in e-waste recycling in the country; there are limited legal e-recyclers, high cost and bureaucracy, informal e-waste handling, few collection points, and limited public awareness [16, 28]. If strategies in East Africa Communication Organization are successful, the sustainable and safe management of e-waste presents will be a major opportunity to turn what was once hazardous waste processing into green jobs. By formalizing e-waste collection, repair, recycling, and processing, countries can create safer working conditions, reduce environmental damage, and stimulate economic growth.

3 Tools and Methodology

We mostly used the SCOPUS database and the Google Scholar web search engine to look for scientific publications (academic articles) in order to conduct this literature evaluation. Of the articles analyzed, 28 were open-access publications that included research on technologies linked to green computing, renewable energy, the circular economy, and the green development action plan. Fifteen papers were eliminated because they were too similar to other studies that were examined or because they didn't seem to be related to the goals of green ICT's. Result classification was based on the commonly used technology solutions driving the circular economy across different industries: (1) communication technologies and green ICT emerging trends (cloud computing, distributed computing, edge computing, renewable energy, artificial intelligence, green data centers, energy-Efficient computing).

ICT and "circular economy" or "green ICT trends" and "circular economy" or "green development action plan" and "climate change" or "renewable energy" and "climate Change" were the search terms used for this paper.

The meta-data that we looked for included "Title," "Abstract," and "Author Keywords." The information gathered took into account English-language journal articles and conference proceedings papers published between 2011 and 2024.The study focused exclusively on open-access papers, ensuring that all included research was freely available to the public.

4 Findings

4.1 Action on Adaptation to Climate Change

Kenya and Rwanda are on the lead of implementing emerging green ICT trends followed by Tanzania. Countries that use remote sensing equipment for climate monitoring and early warning systems are at an advanced stage of complete remote sensor deployment across all industries.

4.2 Action to Reduce GHG Emission

This includes green ICT emerging trends as indicated in Table 2, proper disposal, and end of life for ICT equipment as discussed in the green development plan which is addressed in a number of policies or legislation's that are in existence that aim at protecting the environment and human health. Kenya and Rwanda have a circular economy policy in place but are still undergoing full implementation of ICT-enabled initiatives. Tanzania is governed by a number of laws and regulations that offer an institutional framework for environmentally sustainable management. A review of various policies indicates that in order to address the various challenges and issues of CE, policies tailored to the circular economy are required.

4.3 Action in Mitigation of Climate Change

Table 2 shows examples of ICT-enabled carbon emission reduction projects that are being gradually implemented; these include smart motors, smart systems, smart logistics, and smart grid. The use of solar-powered routers and environmentally friendly connectivity is growing in the aforementioned countries. Due to numerous difficulties and roadblocks, the adoption of ICT-enabled CE initiatives has not yet reached its full potential which fall into the following five categories: societal, institutional, financial, and technological. There is an East Africa Community Organization strategy in place where the selected countries fall under.

5 Conclusion and Recommendation

This paper was set out to map green ICT emerging trends contribution to climate change mitigation and development, with specific focus on circular economy initiatives, green development plans, and the implementation of green ICT emerging trends in selected countries within the Greater Rift Valley. It was found that

Table 2 ICT-enabled technologies

Circular economy (CE) element	Initiative	Sectors in selected countries	Benefit
Communication technologies	Eco-friendly and energy-efficient	5G fixed wireless access (FWA) communication technologies Kenya: Safaricom and Airtel Tanzania: Vodacome, Airtel, Milicom, and Viettel	Adopting novel communication protocols and techniques, the ICT sector can achieve significant resource savings while maintaining high-performance networks Implementation of energy-efficient protocols, virtualization, AI-driven optimization, and adaptive transmission, the ICT industry can significantly lower its carbon footprint while maintaining high-speed, reliable communications
Cloud computing	Cloud-based services and cloud-based streaming services	Safaricom, liquid Telcom, and dimension data have made significant inroads into the East Africa cloud market	Cloud computing plays a crucial role in green ICT by improving energy efficiency, reducing the need for physical infrastructure, and optimizing resource utilization
Distributed computing	Machine-to-machine (M2M) communication	East Africa unified communication, video conferencing, and cloud computing	M2M communications play a crucial role in IoT ecosystems, enabling devices to communicate with minimal human intervention. However, as the number of connected devices grows, energy efficiency becomes a major challenge. To address this, specialized energy-efficient routing protocols have been developed, optimizing power consumption and extending device lifespan

(continued)

Table 2 (continued)

Circular economy (CE) element	Initiative	Sectors in selected countries	Benefit
Edge computing	(Blockchain) thin client	East Africa fintech, humanitarian, agriculture and supply chain, and healthcare	Block chain technology is revolutionizing the growth of community microgrids, which enables local communities to generate, store, and distribute electricity efficiently, often using renewable energy sources like solar and wind
Renewable energy	Use of wind and solar power to power data centers and other IT infrastructure	East Africa renewable energy companies: Voith Gmbh & co. KGaA, Kenya electricity generating company PLC	The cost of renewable energy continues to decrease, making it more economically viable for companies to switch to renewable energy sources
Artificial intelligence	Improve energy efficiency in various sectors, such as transportation, manufacturing, and building management	East Africa healthcare, agriculture, finance, and education	AI can be used to optimize energy usage and reduce waste
Green data centers	Leverage renewable energy sources, such as solar and wind, to power their operations	Kenya: Olkaria Encloud data center	Enhanced sustainability by implementing energy-efficient cooling systems and smart design principles to reduce energy consumption and waste
Energy-efficient computing	Efficient computing systems, such as low power processors	Electric vehicles, tracking, healthcare, consumer electronics, and smart energy	Essential to reducing energy consumption

Source: [5, 30–32] and Author own contribution

emerging trends of green ICT have not been fully exploited, so as to achieve its potential enhancement of sustainable development. Also, there is a need for the countries to work closely with Rwanda which is on the lead in implementing a circular economy. African countries are encouraged to work with stakeholders in green tech revolution and donor funding projects so as to achieve the objectives on implementing ICT CE-enabling initiatives. Further research needs to be done on aligning the green ICT policy, innovation, and investments and ridging the green ICT emerging trends skills gap.

Author's Contribution The author's contributions are as follows: conceptualization: Irine S. Fancy K., Godfrey K., and Lucy K. and methodology: Irine S. Fancy K., Godfrey K., and Lucy K.

Conflict of Interest Statement The authors declare no conflicts of interest.

References

1. Act of Parliament K. (2022). Sustainable waste management act [Internet]. Kenyalaw.org.
2. Africa Ltd S core. (2024). Trends of cloud solutions and IT infrastructure in East Africa for 2023 – Servercore [Internet]. Servercore. [cited 2025 Jan 21]. Available from: https://servercore.com/blog/articles/trends-of-cloud-solutions-and-it-infrastructure-in-east-africa-for-2023/
3. Andreopoulou, Z., Stiakakis, E., & Vlachopoulou, M. (2013a). Green ICT applications towards the achievement of sustainable development. In *E-innovation for sustainable development of rural resources during global economic crisis* (pp. 11–21). IGI Global. https://doi.org/10.4018/978-1-4666-4550-9.ch002
4. Dickerson, K., Torres, D., & Manuel, J. (2011, Feb). *Using ICTs to tackle climate change* [Internet]. Available from: https://www.itu.int/dms_pub/itu-t/oth/0B/11/T0B1100000A3301PDFE.pdf
5. Demestichas, K., & Daskalakis, E. (2020). Information and communication technology solutions for the circular economy. *Sustainability (Switzerland), 12*(18), 1–19. https://doi.org/10.3390/su12187272
6. Evans, O., & Mesagan, E. P. (2022). ICT-trade and pollution in Africa: Do governance and regulation matter? *Journal of Policy Modeling, 44*(3), 511–531.
7. European Parliament. (2023). *Circular economy: Definition, importance and benefits* [Internet]. European Parliament. Available from: https://www.europarl.europa.eu/topics/en/article/20151201STO05603/circular-economy-definition-importance-and-benefits
8. Fata, I. (n.d.). *Solutions based on information and communication technologies for the circular economy*. https://www.researchgate.net/publication/368959085
9. Forti V, Baldé P, Kuehr R, Bel G. The global E-waste monitor 2020.
10. House, C. (2021). *Policies* [Internet]. Circular Economy. Available from: https://circulareconomy.earth/
11. Malo, I. (2024). *Green technology and youth employment in Africa: A transformative opportunity* [Internet]. AFRI. Available from: https://afripoli.org/green-technology-and-youth-employment-in-africa-a-transformative-opportunity
12. *Kenya Innovation Outlook (KIO) EMERGING TECHNOLOGIES EDITION*. (2024). www.kenia.go.ke
13. kictanetadmin. (2024). Greening the internet for East Africa's sustainable digital future | KICTANet think tank [internet]. Kictanet.or.ke. [cited 2025 Jan 20]. Available from: https://www.kictanet.or.ke/greening-the-internet-for-east-africas-sustainable-digital-future/
14. Kushwaha, N. (2863). Green computing: on latest trends. *International Research Journal of Modernization in Engineering*. https://doi.org/10.56726/IRJMETS45745
15. Milenge, D. (n.d.). Circular economy: Africa's perspectives.
16. Nelson, N., Dongjie, N., Mwamlima, P., & Mwitalemi, S. (2021). Assessment of dealers' contribution in current and future E-waste management trends within Dar es Salaam, Tanzania. *International Journal of Scientific Advances, 2*(3).
17. Nyabuto, G. M. (2024). A review of current research trends in green computing. *East African Journal of Information Technology, 7*(1), 51–59. https://doi.org/10.37284/eajit.7.1.1707

18. Osakwe, S., Humeau, E., Bavo, C., Mushi, M., Tricarico, D., Mbugua Hsc, C., et al. (2023 Jan.) Tanzania's Digitalisation Journey: Authors and contributors [Internet]. Available from: https://www.gsma.com/solutions-and-impact/connectivity-for-good/mobile-for-development/wp-content/uploads/2023/01/gsma_ciu_Tanzania_digitalisation_2501_23.pdf
19. Republic of Rwanda. (2022). Revised Green Growth and Climate Resilience National Strategy for Climate Change and Low Carbon Development [Internet]. Available from: https://www.rema.gov.rw/fileadmin/user_upload/Rwanda_Green_Growth___Climate_Resilience_Strategy_06102022.pdf
20. Risola, N. (2024). *Blockchain for agricultural supply chain Management in East Africa Blockchain for agricultural supply chain Management in East Africa Proposed Structure: Blockchain for agricultural supply chain management in East Africa*. https://www.researchgate.net/publication/381305379
21. Rutayisire, J., Markon, S., & Raymond, N. (2017). IoT based coffee quality monitoring and processing system in Rwanda. In *Proceedings of the 2017 IEEE international conference on applied system innovation: Applied system innovation for modern technology, ICASI 2017* (pp. 1209–1212).
22. Rwanda M of I. (2018). *MINICT: Innovation & emerging technologies directorate general* [Internet]. www.minict.gov.rw. Available from: https://www.minict.gov.rw/programs/innovation-emerging-technologies-directorate-general
23. Rwanda National Circular Economy Action Plan and Roadmap January 2022. (n.d.).
24. Simba, A., Ooro, J., & Mbera, O. (2021). *Regional e-waste management strategy 2022–2027 developed by the regional e-waste management steering committee under the leadership of EACO working group 07 on e-waste management and green ICTS* [Internet].
25. Supporting a Sustainable Waste and Circular Economy in Rwanda. (n.d.). www.giz.de
26. Szövetség, M. A., & Andreopoulou, Z. S. (2012). Green informatics: ICT for green and sustainability. *Agrárinformatika/Agricultural Informatics, 3*(2) http://www.magisz.org/journal
27. Tanzania E-Government. (2025). *Government Mobile platform citizens portal* [Internet]. Ega.go.tz. [cited 2025 Jan 20]. Available from: https://mportal.ega.go.tz/
28. The United Republic of Tanzania National E-Waste Statistics Report, 2019 Tanzania Mainland. (n.d.). www.nbs.go.tz
29. Whyte, C. R., Irakoze, G., Katanisa, P., Desmond, P., Hemkhaus, M., Ahlers, J., van Hummelen, S., McGovern, M., & Artola, I. (2020). *Circular economy in the Africa-EU cooperation : Country report for Rwanda*. Publications Office of the European Union.
30. Okele, K., Joiner, J., & Kolta, E. (2022). *Kenya market case study* [Internet]. [cited 2025 Jan 21]. Available from: https://www.gsma.com/solutions-and-impact/connectivity-for-good/mobile-for-development/wp-content/uploads/2023/04/IoT-and-Essential-Utility-Services-Kenya-market-case-study.pdf
31. Okeleke, K., Bororke, S., & Putcha, S. (2023). *5G FWA in Africa emerging trends and opportunities* [Internet]. Available from: https://data.gsmaintelligence.com/api-web/v2/research-file-download?id=79790748&file=151123-5G-FWA-in-Africa.pdf
32. Mbego, S. (2023). *Fully green data centre unveiled in Kenya* [Internet]. CIO Africa. [cited 2025 Jan 21]. Available from: https://cioafrica.co/africa-first-green-data-centre-launched-in-kenya/

Adaptation to Climate Change and Household Food Security: Insight from Kenyan Smallholder Dairy Farmers

Vincent Ngeno

Abstract

In spite of the fact that climate change adaptation plays an important role in improving global food security, little is known about its impact on household food security, particularly for small-scale dairy farmers. Using the food insecurity experience scale (FIES) and the food consumption score (FCS), this study provides insight on how climate change adaptation affects household food security. In South Rift, Kenya, 811 small-scale dairy farmers were interviewed for this cross-sectional study. Two-stage residual inclusion (2SRI) approach was employed to solve the endogeneity issue in the modeling of household food security in response to climate change adaptation. This was achieved by taking into account both observable and unobservable factors. According to the empirical results, climate change adaptation significantly improves household food security. As a result, small-scale dairy farmers should continue to adapt their farming practices to climate change in order to ensure their food security.

Keywords

Climate change adaptation strategies · Smallholder dairy farmers · Two-stage residual inclusion (2SRI) · Food security

V. Ngeno (✉)
Department of Agricultural Economics and Resource Management, Moi University, Eldoret, Kenya

N. Ronoh, C. Rono (eds.), *Rethinking Our Ecosystems*, Springer Proceedings in Geography and Sustainability, https://doi.org/10.1007/978-3-031-95781-9_9

1 Introduction

Climate change is one of the greatest challenges of our time. The evidence now overwhelmingly demonstrates that climate change is unambiguously occurring and at an unprecedented rate [2, 3, 32]. Climate change is also a major problem globally, since temperatures and humidity are constantly rising and altering the living conditions of all species. It is the agriculture sector that is most vulnerable to climatic changes [31, 53] that also affects the livelihoods of farmers. Smallholder farmers in developing countries are particularly vulnerable to climate change [67, 78]. As climate change continues to intensify, in addition to altering pest and disease outbreaks, the frequency and severity of droughts and floods will increase, and the likelihood of poor yields, crop failures, and livestock mortality will increase as well [35, 107].

A large portion of the economy of developing countries is dependent on the agriculture sector, which has been greatly damaged by climate change. Kenya is the 31st most vulnerable country and the 37th least ready for climate change—meaning it is vulnerable to, and unprepared to deal with, the effects of climate change [65]. Models project that climate change will lead to losses equivalent to 2.6% of the gross domestic product (GDP) annually by 2030 [7, 135]. As drought levels increase, staple crops and livelihoods are at risk as well, with 42% of GDP expected to be exposed to drought, versus 16% at present [23, 102]. In many studies conducted in the country, climate change has been associated with significant effects on crops production [77, 116], livestock production [86, 99], and more specifically, dairy farming [25, 87]. The agricultural sector has been shrinking for two decades, but livestock's contribution to agriculture GDP (42%) and country GDP (12%) has been increasing with an average growth rate of 3.5% [103, 108].

In Kenya, almost two thirds of the population (60%) lives below the poverty line and is dependent on agriculture [98, 106]; almost two thirds (60%) work in agriculture, which is mostly rainfed (98%) and is threatened by climate shocks. As a result, there will be a 50% increase in agricultural losses, with cassava, sweet potato, wheat, cabbages, and key exports such as coffee and tea being the worst-affected crops [40]. As a result of drought, livestock will be exposed to two times as much stress [96]; prolonged periods of high temperatures will result in more heat stress, disease prevalence, livestock deaths, and lower production [128]. Kenya is the largest milk producer in Africa with 4.6 million dairy cattle and 5 billion liters of milk produced per year [8, 140]. The government's own dairy policy aims to increase milk production by 130% by 2030, compared with a projection of an 85% increase in milk demand over the next 50 years [21]. Currently, smallholder farmers in Kenya are getting much attention due to intensive government intervention and some policy measures aimed at their survival. The dairy industry is dominated by smallholder farmers, who contribute more than 80% of the livestock production. Several challenges face them, and drastic changes in climatic conditions are among them [16, 79].

The current dairy system in the world is suffering from the effects of climate change. Agricultural productivity, animal health, and animal production are all affected by global warming and the changes it causes to mean climate variables and climate variability. There are also implications for dairy products when it comes to processing, storage, transport, retailing, and consumption caused by climate change. The current dairy systems of smallholder farmers are therefore ineffective at supporting livelihoods and meeting the growing demand for dairy products. A major role is played by the dairy sector in the food supply and food security of the world. In terms of global supply of calories and protein per capita, milk contributes 12% and 27%, respectively, although regional differences exist [24]. The majority of milk produced worldwide comes from grazing systems, on land that is often unsuitable for crop cultivation [63]. A smallholder farmer's livestock can also provide a range of other services to many communities, including a source of energy, a means of transportation, a source of nutrients for poor soils, a source of income generation and diversification, and a form of financial capital. All of these contribute to the overall well-being of communities and contribute to their resilience [23, 124]. About 844 million people earn income from agriculture globally, and dairy contributes about 30% to the agricultural value chain [49, 122]. Even though the full extent and magnitude of climate change's impacts are unclear, dairy sector contributions to food security and sustainability will be affected.

It is a risky activity to engage in dairy farming and to operate in the dairy sector in developing countries. This is especially true in Kenya where dairy enterprises operate under an uncertain and risky situation due to climate change [25, 87]. So, dairy farmers' risk attitudes determine how they adapt to climate change risks. Researchers currently conducting research on climate change show that policy makers are attempting to understand farmers' decisions so as to develop better adaptation policies. It is important for policy makers to understand farmers' behavior better in order to predict how they will react to given policies [95]. It has been shown that there are many studies relating to climate change [4, 57, 64, 77, 80, 82, 93] that focus on crops, while only a small number [20, 30, 54, 56] is conducted in livestock farming. Therefore, dairy farming is severely affected due to climate change in Kenya, and farmers are seriously considering adaptation measures to keep their dairy farms alive. To mitigate the adverse effects of climate change on farm production, dairy farmers used a variety of adaptation measures such as selling weak and diseased animals, migrating during floods and droughts, changing cropping patterns for fodder production, and vaccinations of dairy animals. A variety of adaptation measures are employed across countries based on their respective environments and climates. For example, most farmers in rainfed regions migrate their animals to pastures for feeding and water [91]. Flood-prone areas, too, force farmers to migrate elsewhere in order to procure the feed they need for their dairy animals [89]. As farmers have to incur costs or forgo some of the potential benefits of agricultural production when using adaptation measures, these decisions are not always straightforward [23]. Understanding how these risks are perceived and interpreted locally is necessary to understand the complexity of the adaptation process.

There have been relatively few studies that have discussed the relationship between climate change adaptation and food security [17, 22, 66, 77, 134, 136]. As Kogo et al. [77] explains, climate change will continue to negatively affect crop production and food security in arid and semiarid regions of Kenya, especially for communities that are already vulnerable. As a result of climate variability, cropping patterns and yields are likely to change in several regions. Considering the dramatic growth rate in the country's population and rapid urbanization, crop production and food security systems must adapt to climate variability and change. As per Yan and Alvi [136], climate and weather greatly influence agriculture, especially crop production. As a result of climate change and extreme weather conditions, farmers in South Asia currently face hunger, poverty, and low incomes. Additionally, they are vulnerable to the effects of climate change. Due to the disproportionate impact of climate change on vulnerable populations, they now need to develop climate change adaptation strategies. Food security and climate change adaptation and mitigation require a combination of adaptive and mitigation priorities and long-term approaches. As Islam et al. [66] points out, climate change and climatic extremes adversely affect food security and hamper the country's attainment of the sustainable development goals, specifically goals 1 and 2. This leads to the conclusion that Bangladesh's food security and economy are negatively impacted by climate change.

It has been documented that climate change adaptation and food security are associated, but there are still two major research gaps to be filled. As a result, most research has been focused on crop agriculture sectors [17, 22, 134, 136], which cannot be generalized for dairy sectors since they have different characteristics and demographics. As a result, adaptation may have varying effects on household food security. Abbas et al. [1], Amamou et al. [9], and Kalaugher et al. [70] are the only studies that look at smallholder dairy farmers. Their analyses did not examine whether the adaptation variable was endogenous, and they used propensity score matching. Smallholder dairy farmers, however, decided to adapt to climate change themselves, normally through self-selection based on observable factors such as sociodemographic profile, and unobservable factors such as motivation and skill. Propensity score matching would be invalidated by climate change adaptation since the adaptation is endogenous. Consequently, to obtain more accurate results, endogeneity should be carefully estimated in the context of climate change adaptation. Smallholder dairy farmers are thus targeted in this study in order to estimate the effects of climate change adaptation on their food security. As a result of this study, stakeholders will have new insights for addressing climate change, improving food security, and supporting the Sustainable Development Goals (SDGs), specifically those targeting "climate action" and "no hunger."

As a result of this study, three contributions to the literature have been made. In order to consider the endogeneity issue in the practices, this study uses a two-stage residual inclusion approach (2SRI) and a two-stage predictor substitution approach (2SPS). By investigating the effect of climate change adaptation intensity on dairy farmers' food security, the study reinforces previous research findings. In order to determine the impact on food security of climate change adaptation strategies adopted by smallholder dairy farmers, a variety of strategies were considered. As a last but not least, the study incorporates the food insecurity experience scale (FIES) into the equation, rather than using food consumption scores (FCS) alone as in Hilemelekot et al. [62] and Makate et al. [83]. The FCS represents dietary diversity and nutrient intake of individuals and households [43]. According to FAO [47] and FAO and UNICEF [43], FIES can provide a valuable complement to FCS findings since it measures "food insecurity severity experience matrix" based on immediate answers to questions about access to adequate food.

The following are the sections of this paper: Section 2 discusses an overview of Kenya's dairy industry. A review of the methodology, including research data, measurements of food security, and estimation strategies, is presented in Sect. 3. Presented in Sect. 4 are the results of the research, along with a discussion of the empirical findings and descriptive statistics. Conclusions and policy implications are presented in Sect. 5.

2 Kenya's Dairy Industry: An Overview

It is estimated that Kenya's dairy sector contributes 14% to the country's agricultural GDP. In the dairy industry, smallholder dairy farmers produce 56% of the total milk. About 80% of the sector's producers are smallholder farmers, estimated at 1.8 million. A further 44% of milk is produced by commercial farmers [44]. Among Kenya's dairy production systems, there are three main types. A fully enclosed, zero-grazed intensive system; a semi-intensive system in which animals roam free; and a semi-intensive system in which animals are partly enclosed, zero-grazed, and taken to fields. It is common to find both indigenous and exotic breeds of dairy cattle in Kenya, as well as crossbreed cattle of both varieties. An estimated 4 billion liters of milk are produced annually by more than 5 million dairy cattle with approximately 150% of milk will be produced by 2050 [61].

Kenya is the country with the highest per capita milk consumption in sub-Saharan Africa, with 118 liters consumed annually per person. A growing population is also expected to increase milk demand, which is currently 8 billion liters [81]. Therefore, the government has prioritized this sector in its national strategies and plans, such as the Agriculture Sector Transformation and Growth Strategy (2019–2029) and the president's Bottom-Up Transformational Agenda. The dairy industry will also be guided by a dairy master plan until 2030. Although the sector is capable of reaching its full potential, it faces significant challenges. In order to meet milk demand, Kenya must import it from neighboring countries. Dairy productivity is low, ranging between 6 and 8 liters per day per cow, which accounts for one of the reasons. There is an important distinction to be made between productivity and production systems. In intensive production systems, productivity is highest. When productivity is low, production costs increase and industry competitiveness is affected.

The dairy industry is affected by climate change globally, and its effects are expected to have significant impacts on environmental performance. There is no doubt that dairy is one of the most vulnerable sectors of the food industry to the effects of climate change and growing demand [58]. It is anticipated that climate change will have a predominant impact on the dairy industry in three ways: Firstly, even though global warming is not expected to have adverse effects everywhere, a relevant increase in drought events is predicted across the globe, which will affect crop yield and, as a result, reduce animal feed availability [1]. In addition, rising temperatures may result in cows suffering from heat stress, decreasing their milk production, and increasing their mortality rate [55]. In addition, food safety concerns are exacerbated by the adaptation of food-borne pathogens due to global warming, with a possible alteration in their heat resistance and/or reproduction rates. If this scenario occurs, it is inevitable that food safety measures such as intense heat treatment and/or lower refrigeration temperatures would have to be implemented to maintain food safety [59, 119]. We need to address these challenges across the dairy supply chain before their consequences become unmanageable and negatively impact a safe and adequate supply of milk.

Hence, the dairy sector has a challenge to satisfy future demand for dairy products while reducing its environmental impact and food safety risks under stressful climate conditions. In order to connect vulnerable elements of the dairy value chains with anticipated changes in their performance, it would be necessary to provide evidence and understanding of the dairy sector-climate change interaction. In the near term, this will enhance dairy farmers' and the dairy industry's ability to identify threats and opportunities, allowing them to facilitate more efficient and sustainable decision-making to adapt to climate change.

3 Methodology

3.1 Research Data

Agricultural productivity, food security, and economic development are all threatened by climate change in Kenya. The effects of climate change will increasingly put pressure on milk production, which will undermine food security, lead to micronutrient deficiency, and lead to loss of livelihood because of reduced milk production [28]. In the Kenya highlands, where the agroecological conditions are favorable for dairy farming, model-based estimates have shown that, as temperatures rise in the twenty-first century, milk productivity could decrease by 5%–10% on average [84, 127]. As global warming increases between 1.5 °C and 2 °C, the risk of low milk productivity and food security in vulnerable regions will increase from moderate to high if no or low levels of adaptation are made. At a global warming level of 1.5 °C to 2 °C, droughts, floods, and heat waves will become more frequent, severe, and intense. Several factors contribute to the low milk production of small-scale subsistence dairy farmers, including low adoption of technology and severe weather conditions, including high average temperatures, erratic rainfall patterns, and frequent droughts [34, 115].

Data were collected from small-scale dairy farmers in the South Rift region of Kenya between January and March 2023 in a cross-sectional study. In this study, multistage sampling procedures were used to gather cross-sectional data from the sample population. In the first step, two counties from the South Rift region were selected: Bomet and Kericho, both of which possess medium to high dairying agroecological potential. Smallholder mixed farms account for a considerable portion of milk production in these counties, while medium- and large-scale dairy farms account for a smaller proportion. Sub-counties with the largest smallholder dairy communities were then determined according to the agricultural offices of each county. Accordingly, Sotik and Konoin were selected for Bomet county, while Belgut and Bureti were selected for Kericho county. Following this, ten villages with the largest dairying communities from each sub-county were randomly selected. In order to select respondents, a simple random sampling method was used. As a first step, data on smallholder dairy farmers were collected from local agricultural offices. Afterward, the smallholder dairy farmers were shortlisted by including only those living in select villages, who own and operate farms of about 3–5 acres, and keep 2–5 cows which produce about 10 liters of milk a day despite having the potential of 30 liters. The sample frame included 20–25 smallholder dairy farmers from each village, resulting in a total of 931 respondents. As enumerators in the data collection, postgraduate students at local universities were recruited who could speak English and local languages, including Kipsigis. Prior to collecting the data, they were trained.

A structured questionnaire with six distinct sections was used in this study for face-to-face interviews. As a first step, the questionnaire asked about household characteristics, such as the number of family members, age, education, and dairy farming experience; second, asset ownership, including dairy and nondairy assets; third, smallholder dairy farming activities, inputs, and milk production; fourth, cash flow and social capital; and fifth, perception and understanding of climate change. During this part, respondent's perceptions of climate change were explored by asking them "Have you ever heard of climate change?". A definition would be given by the enumerator as either a confirmation or a correction. As a follow-up question, the perceptions of smallholder dairy farmers about climate change were investigated. Following this, a question was asked: "Did you use an adaptation strategy to deal with the impact of climate change?". They were instructed to select a number of the strategies from a list and indicate which ones they had applied. In addition to the literature review and a discussion with smallholder dairy farmers, these options included changing animal breeds, supplementing forage, and destocking herds. Last but not least, questions are asked about food insecurity (adapted from [42]) to calculate the food insecurity experience scale (FIES), as well as about household food consumption (adapted from World Food Programme guidelines [112]) to calculate the food consumption score (FCS).

3.2 Indicators for Measuring Food Security

As a measure of household food security, the present study used the FIES and FCS, which have previously been validated to estimate individual or household food security in developing countries, including Kenya [42, 43, 112]. Food security is measured by FIES based on individual or household experience. As part of the FIES computation, the FAO has developed eight questions [42]. A household's level of insecurity will be categorized by the FIES measurement into severe insecurity, moderate insecurity, mild insecurity, and secure security. According to the FIES categories, an ordinal variable measuring food security was developed: one for those who are severely insecure, two for those who are moderately insecure, three for those who are mildly insecure, and four for those who are secure. In this study, FCS was used as a second measure of food security. As per World Food Programme [112], FCS is a method of measuring food security based on recalled data about the foods consumed by households that are categorized into eight categories: starches, pulses, vegetables, fruit, meat, dairy products, fats, and sugar. The score of each food group is determined by multiplying the consumption frequency by the score for that food group. This formula can be used to calculate the FCS:

$$\begin{aligned} \text{FCS} = {} & \text{freq\_mainstaple}[2] + \text{freq\_pulses}[3] + \text{freq} \\ & \text{\_vegetables}[1] + \text{freq\_fruit}[1] + \text{freq\_meat}[4] + \text{freq} \\ & \text{\_dairy}[4] + \text{freq\_fats}[0.5] + \text{freq\_sugar0.5} \end{aligned} \tag{1}$$

A household's FCS score of 0 to 28 will be considered poor, a score of 28.5 to 42 will be considered borderline, and a score over 42 will be considered acceptable. Using the new scale, create an ordinal FCS variable: one for poor food security, two for borderline food security, and three for acceptable food security.

4 Estimation Strategies

4.1 Modeling How Adaptation to Climate Change Impacts Household Food Security

In order to evaluate the impact of adapting to climate change on household food security, the following model is used:

$$\text{FS}_i^j = \partial_{\text{i}}\text{CC}_{\text{i}} + \varphi_{\text{i}}\text{CV}_{\text{i}} + \mu_{\text{i}} \tag{2}$$

Based on ordinal values, FS_i indicates household food security of smallholder dairy farmers, including food security j = 1 and FCS j = 2, both of which are measured by ordinal values, which were previously discussed. The CC_i treatment variable in this study represents the adaptation to climate change, measured as a dichotomous variable (either 1 if the smallholder dairy farmers are adapting to climate change, or 0 if they are not). The CV_i variable is a vector of the profiles of smallholder dairy farmers, which includes their age, education, family size, assets, and climate perceptions, along with other variables related to smallholder dairy farmers, including food security and FCS [26, 113, 114]. The parameters i and i are to be computed, while i is a random error term.

There may be unobservable variables that influence adaptation to climate change and food security, such as smallholder dairy farmers' motivation and skills; the adaptation to climate change variable CCi is therefore likely to be endogenous in Eq. 2. It is important not to overlook this factor when estimating household food security effects associated with climate change.

To eliminate the endogeneity issue from food security estimation, previous studies identified two effective approaches: the two-stage residual inclusion approach (2SRI) [51, 52, 126] and the two-stage predictor substitution approach (2SPS) [27, 29, 132]. In 2SRI, two steps are involved. In the first step, the treatment variable, which is adaptation to climate change, which is potentially endogenous, is regressed as a function of smallholder dairy farmer profiles, which serve as control variables, and at least one instrumental variable, and a residual is then predicted. The second step involves the estimation of household food security, which includes food security and FCS, using a set of previously mentioned control variables and one additional variable (the residual term from the first step). In contrast to the 2SRI approach, the 2SPS approach predicts an endogenous variable, adaptation to climate change. In the second stage, household food security is regressed as a function of adaptation to climate change, replacing the endogenous variable with the predicted value. In this stage, the same control variables are used as in the previous stage. Consequently, 2SRI and 2SPS are employed in this study. In previous studies, 2SRI has been found to be more effective than 2SPS (e.g., [18, 27, 137, 142]). As a result, this study employed the 2SRI method.

4.2 The Two-Stage Residual Inclusion (2SRI) Technique

The first stage of the 2SRI, adaptation to climate change, is regressed as a function of the smallholder dairy farmer's basic profiles, which are the control variables in Eq. 2, and includes an instrumental variable. Due to the fact that adaptation to climate change is a dummy variable, the first stage of the 2SRI was estimated using a probit model, as follows:

$$\mathrm{CC}_{\mathrm{i}}^{*} = \upsilon_{\mathrm{i}}\mathrm{CV}_{\mathrm{i}} + \omega_{\mathrm{i}}\mathrm{IV}_{\mathrm{i}} + \epsilon_{\mathrm{i}} \text{ with } \mathrm{CC}_{\mathrm{i}} = \begin{pmatrix} 1 \text{ if } \mathrm{CC}_{\mathrm{i}}^{*} > 0 \\ 0 \text{ otherwise} \end{pmatrix} \tag{3}$$

where CC_i* is an observed dummy variable that indicates the probability of the smallholder dairy farmer adapting to climate change, a value of 1 indicates that the smallholder dairy farmer applies the adaptation, and 0 indicates that he or she does not; CV_i describes smallholder dairy farmers' basic profiles, and they are the control variables; in this case, the instrumental variable is IVi; i and i are the parameters that need to be estimated; and *i* represents the random error term. As instrumental variables, Eq. 3 uses perception and information about climate change. Climate change perception and climate information are supposed to impact adaptation to climate change variables, but not on food security and FCS, which are used as household food security indicators. This study employed a Pearson correlation estimation like Fentie and Beyene [46], Kerr et al. [73], and Zahid et al. [138] to determine the validity and suitability of climate change perception and climate information as its instrumental variables. As shown in Table A1, the Pearson correlation test indicates that both climate perception variables and climate information positively and significantly affect adaptation to climate change. In addition, the Pearson correlation results also indicate that all instrumental variables do not have a significant effect on the outcome variables (food security and FCS), suggesting that climate change perception and climate information are valid instrumental variables.

Using an ordered probit model, a residual term that has been predicted in Eq. 3 is included as another control variable in the second stage of 2SRI estimation. As a result, the second stage model is specified as follows:

$$\mathrm{FS}_{\mathrm{j}}^{\mathrm{i}} = \alpha_{\mathrm{i}}\mathrm{CC}_{\mathrm{i}} + \beta_{\mathrm{i}}\mathrm{CV}_{\mathrm{i}} + \gamma_{\mathrm{i}}\mathrm{RT}_{\mathrm{i}} + \nu_{\mathrm{i}} \tag{4}$$

where FS_ij, CC_i, and CV_i are previously discussed variables and RT_i represents the residual term predicted from Eq. 3 after estimation. Specifically, Terza [126] argue that if the RT_i significantly affects the study's outcome variables of food security and FCS, and then it implies that the unobservable variables could provide a bias in Eq. 2. This implies that the 2SRI model is the most appropriate one for estimation. α_i, β_i, and γ_i correspond to parameters to be estimated, and i refers to the random error term for Eq. 4.

5 Result and Discussion

5.1 Descriptive Statistics

Based on descriptive statistics and measurement of the selected variables, Table 1 indicates approximately 61.3% of smallholder dairy farmers have employed climate change adaptation strategies, with an adaptation intensity of 3.17, indicating at least three combinations of climate change adaptation strategies have been employed by smallholder dairy farmers. In order to adapt to climate change, smallholder dairy farmers have implemented following major strategies: improved feeding, the provision of medical facilities, the use of seasonal and weather forecast information, livestock diversification and stress-tolerant breeds and species, and the development of improved/stress-tolerant livestock.

Based on the results of the study, smallholder dairy farmers have adopted an improved feeding strategy composed of grazing management, supplemental feeding, concentrate feeding, and bran feeding, with approximately 55.14% adopting the strategy. With a 34.6% adaptation rate, the second most common climate change adaptation strategy was the provision of medical facilities, including disease control precautions, livestock training, drinking water, and sanitation services. Also, 28.6% of smallholder dairy farmers adopted seasonal and weather forecast updates as a third climate change adaptation strategy. With regard to climate change adaptation strategies, smallholder dairy farmers were not much into livestock diversification and improved/stress-tolerant breeds/species with a 15.7% adoption rate.

Moreover, according to study results, the food security and FCS variables indicated the level of food security, averaging 3.131 and 2.344, respectively. Based on these values, the smallholder dairy farmers in the study have a food security status equal to 'mildly insecure' based on the food security variable, and "borderline insecure" based on the FCS variable.

Table 1 Variable descriptive statistics

Variables	Measurement	Mean	Standard Deviation
Treatment variables			
Adaptation	If the farmer adapts to climate change, 1; otherwise, 0	0.613	0.488
Adaptation intensity	0 to 4 adaptation strategies	3.17	1.105
Outcome variables			
Food security	1 = severely insecure, 2 = moderately insecure, 3 = mildly insecure, and 4 = food secure	3.131	1.113
FCS	Poor is scored 1, borderline is scored 2, and acceptable is scored 3	2.344	0.668
Independent variable			
Age	Smallholder farmer's age in years	43.536	10.083
Education	Smallholder farmer's level of education in years	8.268	3.903
Experience	Smallholder farmer's dairy experience in years	24.798	8.167
Household size	Members of the family	6.113	1.699
Land size	Acreage under dairy production	2.106	1.552
Off-farm income	If household members do off-farm work, 1; otherwise, 0	0.371	0.449
Extension	1 if farmer benefited from government extension agent, 0 otherwise	0.216	0.194
Credit	0 if the farmer lacks access to credit, 1 if farmer does	0.221	0.116
Membership	Farmer group membership: 1 = member, 0 = not a member	0.44	0.25
Instrumental variables			
Rainfall	Farmer's perception of rain intensity ranges from 1 = very decreased to 5 = significantly increased	3.107	1.117
Temperature	Temperature perception by farmers ranges from 1 = very low to 5 = significantly high	3.413	0.882
Information	If the farmer has access to climate information, 1; otherwise, 0	0.536	0.383

For dummies, the mean refers to the mean proportions in percentages

Household heads had an average age of 44 years and 25 years of farming experience, as well as about 8 years of formal education, indicating that respondents were middle-aged with considerable farming experience and a good level of formal education in the study area. Families averaged 6.113 members per household. Milk production relies heavily on land; smallholder dairy farmers in the study area owned about 2.106 acres of land on average. Around 37% of smallholder dairy farmers also engage in off-farm work to supplement income from on-farm activities. To buffer farm income, smallholder dairy farmers engaged in off-farm income during the dry season every year. The technical advice of agricultural extension services had been helpful to about 22% of household heads regarding dairy production, whereas 78% of smallholder dairy farmers did not get any help from extension services. Due to an inadequate number of extension agents, smallholder dairy farmers have limited interactions with agricultural extension services agents. Smallholder dairy farmers produce dairy in small areas due to limited access to credit for farm expansion, and 22% of smallholder dairy farmers use credit for dairy production, indicating that credit is limited to smallholder dairy farmers. Household heads belonging to farmer organizations comprised only 44% of the households.

Based on the mean differences in variables between adapters and non-adapters, Table 2 shows the mean differences in variables between them. A systematic difference is observed between adapters and non-adapters, as indicated by the results. Climate change adapters and non-adapters have significant differences in variables including their education years, income from off-farm sources, access to extension services, credit services, membership in farmers associations, and perception of rainfall intensity. In terms of education, climate change adapters are more educated than non-adapters, with average years of education varying significantly between the two groups. It's likely that farmers who are more educated are more aware of climate change, adaptation practices, and the benefits that can be derived from these practices [6]. As a result of supplementary income from off-farm activities, climate change non-adapters did better in accessing off-farm income than climate change adapters, contrary to the belief that supplementary income from off-farm activities influenced climate change adaptation strategies. Additionally, farmers engaged in off-farm activities typically had more financial resources to adapt to climate change, but did not.

Adapters and non-adapters have significant differences in access to agricultural extension services, so information provided by extension agents assists farmers in making informed decisions about incorporating new innovations, including climate change adaptation [74]. Access to credit facilities differs greatly between adapters and non-adapters; agricultural credit increases households' productive and physical

Table 2 Climate change adaptation mean differences for selected variables

Variables	Adaptation	Non-adaptation	Mean differences
Food security	3.341	2.934	0.407***
FCS	2.533	2.148	0.385***
Age	42.448	44.627	−2.179
Education	8.845	7.689	1.156**
Experience	24.209	25.387	−1.178
Household size	5.485	6.741	−1.256
Land size	2.439	1.773	0.666
Off-farm income	0.229	0.513	−0.284***
Extension	0.267	0.165	0.102*
Credit	0.313	0.129	0.184***
Membership	0.565	0.329	0.236***
Rainfall	3.869	2.345	1.524***
Temperature	3.705	3.121	0.584
Information	0.643	0.429	0.214

Note: *, **, and *** indicate significance for 10%, 5%, and 1%, respectively

capital levels. There are many factors that limit agricultural investments, but credit constraints, credit market imperfections, and restrictive credit accessibility conditions play a significant role in making agriculture, particularly climate change adaptation strategies, inherently unattractive [105]. A significant difference exists between adapters and non-adapters in their membership to agricultural organizations; Khanal et al. [75] suggest that farmers' adoption of climate change adaptation is enhanced by membership in agricultural organizations by influencing the members of these organizations to switch from traditional agricultural practices to practices that are better suited to climate change. In terms of perceptions of rainfall change, adapters and non-adapters differ significantly. According to Sutcliffe et al. [125], farmers' perceptions of changes in rainfall may affect their perceptions of climate impacts' seriousness, which may influence whether or not adaptations are implemented. As shown in Table 2, the adapters scored higher than the non-adapters based on their food security and FCS variables. This result, however, cannot be used to claim that adaptation to climate change has a positive impact on food security for smallholder dairy farmers. Based on the results in Table 2, it is not clear what factors influence smallholder dairy farmers' decisions to adapt to climate change and improve their food security.

6 Empirical Results and Discussions

Throughout this section, the 2SRI results are discussed, beginning with the first stage results, which highlight the factors that influence the adoption of climate change strategies, and then the second stage results, which examine the impact of adaptation and control variables on household food security. A final estimation is presented on how adaptation intensity affects household food security.

Smallholder dairy farmers' age is negatively correlated with their adoption of climate change adaptation strategies. Smallholder dairy farmers, especially young ones, are more likely than their older counterparts to adapt to climate change because they are innovative and keen to try new technologies and adaptation strategies. There is also a possibility that older dairy farmers are either unfamiliar with recent innovations in climate change adaptation strategies, or they are reluctant to adopt new climate change adaptation strategies. The results of Azadi et al. [15] study have also found that farmers' age has a negative effect on their adoption of coping strategies for climate change.

Climate change adaptation strategies adopted by smallholder dairy farmers are positively associated with their participation in non-farm income activities. Adaptation strategies are more likely to be adopted by smallholder dairy farmers with off-farm incomes, based on the significantly positive relationship between the two variables. By engaging in off-farm activities, small-scale dairy farmers can purchase improved feeds, medical services, and invest in seasonal and weather forecasting information, as well as livestock diversification, stress-tolerant breeds and species, and development of improved/stress-tolerant livestock, as their credit constraints may be overcome through their involvement in off-farm activities. Climate change adaptation is increased by participation in non-farm income activities, according to Amfo and Ali [10] and He et al. [60].

Smallholder dairy farmers with formal education are more likely to adopt climate change adaptation strategies than those with no formal education, as indicated by the coefficient of education's positive sign and statistical significance. A smallholder dairy farmer who is educated is more likely to be knowledgeable about climate change and related issues. Access to information about improved technologies and higher productivity are believed to be associated with higher education levels [92]. Based on a number of studies, there appears to be a positive association between a household head's education level and the adoption of new technologies and adaptation to climate change [130]. The likelihood of smallholder dairy farmers adapting to climate change is therefore greater when they have higher levels of education. Assimilation of information, the ability to adopt new technologies, and the ability to anticipate changes are all enhanced through education, which will increase their likelihood of adopting different adaptation strategies. Previous studies by Feinstein and Mach [45] and Wamsler et al. [131] support the results of this study.

Contrary to expectations, smallholder dairy farmers' years of farming experience had a negative association with their adoption of climate change adaptation strategies. A smallholder dairy farmer's farming experience may decrease their likelihood to adopt climate change adaptation strategies, resulting in a decrease in the likelihood of farmers adopting climate change adaptation strategies. According to previous studies such as Aryal et al. [13], the negative sign may be due to the age of highly experienced farmers.

Compared to younger dairy farmers desperate to gain experience, older smallholder dairy farmers with high experience are less likely to adopt the mentioned strategies. A similar conclusion was reached by Elahi et al. [41], who found that age was negatively associated with adoption of climate change adaptation strategies among smallholder dairy farmers, including buying improved feed and breeding stress-tolerant dairy varieties. A farmer's age can be thought to be related to their level of experience, which is important to their likelihood of embracing a particular strategy to cope with climate change, according to Donatti et al. [35] and Partey et al. [111].

Smallholder dairy farmers' access to extension services was positive and significantly associated with climate adaptation strategies. Adaptation to climate change can be supported in three ways by extension officers, according to Antwi-Agyei and Stringer [12]: (i) through facilitating and implementing policies, programs, (ii) through providing information about and guiding the management of new farming methods, and (iii) enhancing capacity development. The existing literature indicates that agricultural extension services can enhance farmers' capacity to address climate change effects through educational programs [19]. Atube et al. [14] reported that extension services helped vulnerable individuals and communities build their resilience and adaptive capacities in managing climate change effects by generating awareness about the effects of climate change [141]. Farmer-field schools, flyers, radio messages, and field demonstrations are all methods of providing extension education to farmers around the world.

Credit access has a significant positive impact on smallholder dairy farmers' adoption of climate change adaptation strategies, according to the results. Adaptation strategies to climate change might be capital-intensive, requiring dairy farmers to invest in environmentally sustainable adaptation approaches. In the absence of credit, smallholder dairy farmers might not be able to adopt climate change adaptation strategies, even if they are informed about it [5]. The financial support that smallholder dairy farmers are able to access heavily influences their decision to adopt adaptation strategies, according to Mihiretu et al. [91]. A number of studies, including Antwi-Agyei and Stringer [12], Gezie [53], Jin et al. [68], and Sertse et al. [120], have found the same. Credit provides farmers with the financial capability to cope with the transaction costs related to different climate change adaptation options.

Climate change adaptation strategies are more likely to be adopted by smallholder dairy farmers if they receive climate change information through radio, television, mobile phones, etc. This is because access to climate change information has a positive and significant effect on climate change. Thus, smallholder dairy farmers have an increased ability to adapt to climate change if they are informed about the adverse effects. In their studies on smallholder farming households, Dunn et al. [37] and Owusu et al. [109] argued that climate information on climate and weather conditions, as well as extreme events, can help reinforce adaptation strategies and enhance their planning, implementation, and adoption. It has been demonstrated by Dorward et al. [36] that climate services that take users' motives into account when designing and implementing them often have significant impacts on their livelihoods. Thus, Singh [121] suggest that by increasing farmers' knowledge of climate change adaptation measures and removing barriers to accessing climate information, climate change adaptation measures could be scaled up.

As members of dairy farmer groups or cooperatives, smallholder dairy farmers get access to good inputs, labor, credit, and marketing expertise. As a result, members of local institutions discuss farming problems and explore new opportunities, share experiences, and synthesize new information and innovations. As well as providing information and consultation to members, cooperatives are also capable of increasing access to government funding. In addition, they provide market information and provide technical assistance. This study indicates that climate change adaptation is positively correlated with membership in farmers groups/cooperatives, confirming the findings of studies Bedeke et al. [19], Mwinkom et al. [94], and Zamasiya et al. [139]. Accordingly, policy decisions should take into account the importance of sensitizing local farmers to join local farm associations and groups. Institutions play a key role in reversing the vulnerability of smallholder dairy farmers to climate change and variability, as expressed in the sustainable livelihood framework in climate change adaptations.

Climate change adaptation strategies were significantly influenced by smallholder dairy farmers' perceptions of temperature increase. Climate change adaptation strategies are more likely to be adopted by 6% if temperatures rise 1 °C above average. Reduced water availability, high evaporation rates, and reduced agricultural productivity were associated with prolonged drought. A number of adaptation measures will be used by smallholder dairy farmers in order to cope with increased temperatures, including selling weak or diseased animals, migrating during floods and droughts, changing cropping patterns for fodder production, and vaccinating dairy animals regularly to mitigate climate change's adverse effects. In a similar fashion, Galappaththi et al. [50], Masuda et al. [85], and Niles et al. [101] found that farmers living in relatively high-temperature regions were more likely to adapt to climate change.

Climate change adaptation strategies adopted by smallholder dairy farmers were significantly influenced by their perceptions of rain intensity increases. Due to this, smallholder dairy farmers are more likely to adopt adaptation strategies as rainfall increases. Farmers may be attempting to mitigate climate change's effects due to a greater incidence of floods. In addition to this result, Panda [110], Sutcliffe et al. [125], and Van Der Pol et al. [129] found that increased rainfall intensity increases the probability of smallholder dairy farmers adopting climate change adaptation strategies.

7 2SRI's Second Stage

As illustrated in Table 3, the results of the second stage of the 2SRI approach, estimated by Eq. 4, are shown, namely, the impact of adaptations and control variables on household food security measurements, i.e., food security and FCS variables. The residual variable, which is tested in the second stage of 2SRI, illustrates the endogeneity issue. It confirms endogeneity in the unobservable variables that the residual variable has a positive and significant impact on food security. In addition, it affirms that the 2SRI approach can be appropriate for estimating the impacts of climate change adaptation on household food security.

In Table 3, the impact of climate change adaptation on food security is shown. Food security and FCS are positively and significantly affected by the adaptation according to the results of the 2SRI approach. With these arguments in mind, we may be able to conclude that household food security is significantly impacted by climate change adaptation. Because climate change has major implications for food security and livelihoods in Kenya, those who adopted climate change strategies have a more secure food status than those who have not adopted it. Currently, extreme droughts impede people's ability to grow food and raise livestock, and pastoralists and agro-pastoralists must adapt to climate changes to remain healthy and secure [97, 133]. It is consistent with the results of Leisner [80] and Zavala Nacul and Revoredo-Giha [140], who found an association between climate change adaptation and food security. In this study, however, the novelty is that the endogeneity and adaptation intensity issues are addressed with a 2SRI approach. Researchers have examined the impact of climate change adaptation on food security using dummy variables only [33, 73, 77, 104, 117, 118].

As indicated by the positive and significant coefficient of the variable representing education, higher levels of education may correlate with greater levels of food security, as measured by FIES and FCS. It is consistent with previous studies

Table 3 Estimates of first and second stage 2SRIs

	First stage		Second stage			
	Adaptation		Food security		FCS	
Variables	Coef.	Std. Err.	Coef.	Std. Err.	Coef.	Std. Err.
Adaptation			0.098**	0.047	0.094**	0.041
Age	−0.046**	0.021	0.012	0.010	0.001	0.009
Education	0.029***	0.011	0.034***	0.013	0.043***	0.015
Experience	−0.181**	0.079	0.038	0.029	0.044	0.027
Household size	0.017	0.013	0.021	0.013	0.014	0.011
Land size	0.014	0.011	0.076	0.047	0.061	0.038
Off-farm income	0.035***	0.011	0.023	0.017	0.023	0.015
Extension	0.061***	0.021	0.044	0.027	0.051	0.034
Credit	0.077**	0.032	0.053	0.037	0.078	0.049
Membership	0.1164***	0.044	0.063	0.043	0.081	0.051
Temperature	0.059***	0.018	−0.079***	0.029	−0.029***	0.011
Rainfall	0.048***	0.013	−0.051**	0.021	−0.048***	0.017
Information	0.101**	0.041	0.021	0.015	0.024	0.017
RT			0.132***	0.027	0.128**	0.055
Constant	1.136***	0.021	1.103***	0.043	1.003***	0.089
Cut1	0.101	0.098	0.101	0.098	−0.099	0.061
Cut2	0.113	0.073	0.113	0.073	0.145	0.089
Cut3	0.128	0.084	0.128	0.084		
Log likelihood	−101.104		−191.362		−117.657	
LR chi2(15)	139.169		123.41		163.66	
Prob. > chi2	0.000		0.000		0.000	
Pseudo R2	0.280		0.186		0.157	

Note: The asterisks ***, **, and * represent 1%, 5%, and 10% significance levels, respectively

showing a positive and significant relationship between education and household food security [11, 117, 136]. Farmers may have enhanced food security knowledge due to education. The coefficient of off-farm work is positive for FIES and FCS and statistically different from zero, suggesting that engaging in off-farm work may lead to an increase in food security. The result here is probably due to the income effect of off-farm work participation, with households using earnings from off-farm work to improve their food and nutrition security, a finding that is consistent with earlier research linking off-farm work with food and nutrition security [38, 39, 48, 72].

In terms of food security, the coefficient of smallholder dairy farmers' perceptions of temperature rise is negative for FIES and FCS, indicating that rising temperatures are a limiting factor affecting food security. High temperatures cause reduced crop and livestock yields, including milk quality among smallholder dairy farmers, because they reduce water use efficiency, crop growth period, photosynthesis, and yields. Adaptation is necessary to enhance food security given the changing climate will further aggravate food security challenges. The findings are consistent with those of earlier studies [69, 71, 123] that found a positive and significant correlation between rising temperatures and household food security. Food security may be negatively affected by temperature rise, because it negatively impacts crop and animal production and productivity.

According to the results, smallholder dairy farmers' perceptions of rainfall intensity negatively affect food security as measured by FIES and FCS. As a result, crop and livestock yields have been significantly reduced due to the high variation of rainfall due to the change in climate. Farmers in Kenya and other developing regions rely heavily on rainfed agriculture, so climate is an important resource for both crop and livestock production. Additionally, droughts, changes in rainfall timing, duration, intensity, and geographical location, as well as changes in planting season start or end, may not be suitable for many crops, which could adversely affect households' food security, according to previous studies [76, 88, 90, 100]. Climate change has a multidimensional effect, affecting crop and livestock yields directly. In consequence, these reductions have a direct impact on the quality and availability of farm-produced food. Therefore, households must consume less preferred foods with poor nutritional qualities and contents due to low crop and livestock productivity.

8 Conclusions and Policy Implications

Climate change is becoming a reality today. The change in climate has a substantial effect on food security, and adaptation strategies must be developed to deal with these changes. Among Kenyan smallholder dairy farmers, this study examines the relationship between climate change adaptation and household food security. For addressing the potential endogeneity issue, the study used two-stage residual inclusion (2SRI). As a result of the 2SRI's first stage, empirical results showed that smallholder dairy farmers' ages, education, access to extension services, credit access, climate change information, membership in farmers groups/cooperatives, and perceptions of temperature increase and increased rain intensity influenced smallholder dairy farmers' decisions to adapt to climate change. The 2SRI's second stage, on the other hand, found that climate change adaptation strategies and dairy farmer education had a significant and positive impact on food security, as measured by FCS and FIES, while the perception of increased temperature and increased rainfall intensity negatively impacted food security for smallholder dairy farmers.

Thus, to improve the adaptive capacity of smallholder dairy farmers both at the study area and other regions, it is therefore recommended that policies that would improve the decisions of smallholder farmers to adopt climate change adaptation strategies are put in place. These policies and investment strategies of the government should be formulated and targeted at providing education, credit facilities, and climate change adaptation strategy information. Further, the government and stakeholders should cooperate to improve the efficiency of smallholder dairy extension through the use of dairy extension agents; smallholder dairy farmers should be made aware of climate change by increasing access to climate information, improving conditions under which farmers can access climate change information and appropriate agricultural credit, and providing policy incentives for ensuring the sector's sustainability. Moreover, the findings suggest that smallholder dairy farmers can adapt to climate change as part of efforts to achieve the UN SDGs of ending hunger, achieving food security, and promoting sustainable agriculture.

References

1. Abbas, Q., Han, J., Bakhsh, K., Ullah, R., Kousar, R., Adeel, A., & Akhtar, A. (2022). Adaptation to climate change risks among dairy farmers in Punjab, Pakistan. *Land Use Policy, 119*, 106184.
2. Abram, N. J., Henley, B. J., Sen Gupta, A., Lippmann, T. J., Clarke, H., Dowdy, A. J., Sharples, J. J., Nolan, R. H., Zhang, T., & Wooster, M. J. (2021). Connections of climate change and variability to large and extreme forest fires in southeast Australia. *Communications Earth & Environment, 2*(1), 8.
3. Adak, S., Mandal, N., Mukhopadhyay, A., Maity, P. P., & Sen, S. (2023). Current state and prediction of future global climate change and variability in terms of CO_2 levels and temperature. In *Enhancing resilience of dryland agriculture under changing climate: Interdisciplinary and convergence approaches* (pp. 15–43). Springer.
4. Ahsan, F., Chandio, A. A., & Fang, W. (2020). Climate change impacts on cereal crops production in Pakistan: Evidence from cointegration analysis. *International Journal of Climate Change Strategies and Management, 12*(2), 257–269.
5. Al-Amin, A. A., Akhter, T., Islam, A. H. M. S., Jahan, H., Hossain, M. J., Prodhan, M. M. H., Mainuddin, M., & Kirby, M. (2019). An intra-household analysis of farmers' perceptions of and adaptation to climate change impacts: Empirical evidence from drought prone zones of Bangladesh. *Climatic Change, 156*, 545–565.
6. Ali, A., & Erenstein, O. (2017). Assessing farmer use of climate change adaptation practices and impacts on food security and poverty in Pakistan. *Climate Risk Management, 16*, 183–194.
7. Aligishiev, M. Z., Massetti, E., & Bellon, M. M. (2022). *Macro-fiscal implications of adaptation to climate change*. International Monetary Fund.
8. Alonso, S., Angel, M. D., Muunda, E., Kilonzi, E., Palloni, G., Grace, D., & Leroy, J. L. (2023). Consumer demand for milk and the informal dairy sector amidst COVID-19 in Nairobi, Kenya. *Current Developments in Nutrition, 7*(4), 100058.
9. Amamou, H., Sassi, M. B., Aouadi, H., Khemiri, H., Mahouachi, M., Beckers, Y., & Hammami, H. (2018). Climate change-related risks and adaptation strategies as perceived in dairy cattle farming systems in Tunisia. *Climate Risk Management, 20*, 38–49.
10. Amfo, B., & Ali, E. B. (2020). Climate change coping and adaptation strategies: How do cocoa farmers in Ghana diversify farm income? *Forest Policy and Economics, 119*, 102265.
11. Andrianarison, F. (2022). Unravelling the linkage between food security, poverty reduction, and education for sustainable development. *The Journal of Development Studies, 58*(11), 2198–2221.
12. Antwi-Agyei, P., & Stringer, L. C. (2021). Improving the effectiveness of agricultural extension services in supporting farmers to adapt to climate change: Insights from northeastern Ghana. *Climate Risk Management, 32*, 100304.
13. Aryal, J. P., Sapkota, T. B., Khurana, R., Khatri-Chhetri, A., Rahut, D. B., & Jat, M. L. (2020). Climate change and agriculture in South Asia: Adaptation options in smallholder production systems. *Environment, Development and Sustainability, 22*(6), 5045–5075.
14. Atube, F., Malinga, G. M., Nyeko, M., Okello, D. M., Alarakol, S. P., & Okello-Uma, I. (2021). Determinants of smallholder farmers' adaptation strategies to the effects of climate change: Evidence from northern Uganda. *Agriculture & Food Security, 10*(1), 1–14.
15. Azadi, Y., Yazdanpanah, M., & Mahmoudi, H. (2019). Understanding smallholder farmers' adaptation behaviors through climate change beliefs, risk perception, trust, and psychologi-

cal distance: Evidence from wheat growers in Iran. *Journal of Environmental Management, 250*, 109456.
16. Bagath, M., Krishnan, G., Devaraj, C., Rashamol, V. P., Pragna, P., Lees, A. M., & Sejian, V. (2019). The impact of heat stress on the immune system in dairy cattle: A review. *Research in Veterinary Science, 126*, 94–102.
17. Bastas, K. K. (2022). Impact of climate change on food security and plant disease. In *Microbial biocontrol: Food security and post harvest management* (Vol. 2, pp. 1–22). Springer.
18. Basu, P., Galiè, A., & Baltenweck, I. (2019). Presence and property: Gendered perspectives on participation in a dairy development program in Kenya and Uganda. *Women's Studies International Forum, 74*, 68–76.
19. Bedeke, S., Vanhove, W., Gezahegn, M., Natarajan, K., & Van Damme, P. (2019). Adoption of climate change adaptation strategies by maize-dependent smallholders in Ethiopia. *NJAS-Wageningen Journal of Life Sciences, 88*, 96–104.
20. Bernabucci, U. (2019). Climate change: Impact on livestock and how can we adapt. In *Animal frontiers* (Vol. 9, pp. 3–5). Oxford University Press.
21. Blackmore, E., Guarin, A., Vorley, W., Alonso, S., & Grace, D. (2022). Kenya's informal milk markets and the regulation–reality gap. *Development Policy Review, 40*(3), e12581.
22. Bocchiola, D., Brunetti, L., Soncini, A., Polinelli, F., & Gianinetto, M. (2019). Impact of climate change on agricultural productivity and food security in the Himalayas: A case study in Nepal. *Agricultural Systems, 171*, 113–125.
23. Bogale, G. A., & Erena, Z. B. (2022). Drought vulnerability and impacts of climate change on livestock production and productivity in different agro-Ecological zones of Ethiopia. *Journal of Applied Animal Research, 50*(1), 471–489.
24. Boland, M., & Hill, J. (2020). World supply of food and the role of dairy protein. In *Milk Proteins* (pp. 1–19). Elsevier.
25. Brandt, P., Yesuf, G., Herold, M., & Rufino, M. C. (2020). Intensification of dairy production can increase the GHG mitigation potential of the land use sector in East Africa. *Global Change Biology, 26*(2), 568–585.
26. Brüssow, K., Faße, A., & Grote, U. (2017). Implications of climate-smart strategy adoption by farm households for food security in Tanzania. *Food Security, 9*, 1203–1218.
27. Cai, B., Small, D. S., & Have, T. R. T. (2011). Two-stage instrumental variable methods for estimating the causal odds ratio: Analysis of bias. *Statistics in Medicine, 30*(15), 1809–1824.
28. Change, I. P. on C. (2022). *Climate change and land: An IPCC special report on climate change, desertification, land degradation, sustainable land management, food security, and greenhouse gas fluxes in terrestrial ecosystems*. Cambridge University Press.
29. Chapman, C. G., & Brooks, J. M. (2016). Treatment effect estimation using nonlinear two-stage instrumental variable estimators: Another cautionary note. *Health Services Research, 51*(6), 2375–2394.
30. Chowdhury, Q. M. K., Hossain, M., Ahmed, J., Shykat, C. A., Islam, M. S., & Hasan, M. (2019). Impact of climate change on livestock in Bangladesh: A review of what we know and what we need to know. *American Journal of Agricultural Science, Engineering, and Technology, 3*(1), 89–96.
31. Conway, D., Nicholls, R. J., Brown, S., Tebboth, M. G., Adger, W. N., Ahmad, B., Biemans, H., Crick, F., Lutz, A. F., & De Campos, R. S. (2019). The need for bottom-up assessments of climate risks and adaptation in climate-sensitive regions. *Nature Climate Change, 9*(7), 503–511.
32. Dey, R., & Lewis, S. C. (2021). Natural disasters linked to climate change. In *The impacts of climate change* (pp. 177–193). Elsevier.
33. Diallo, A., Donkor, E., & Owusu, V. (2020). Climate change adaptation strategies, productivity and sustainable food security in southern Mali. *Climatic Change, 159*(3), 309–327.
34. Diniso, Y. S., Zhou, L., & Jaja, I. F. (2022). Dairy farmers' knowledge and perception of climate change in the Eastern Cape province, South Africa. *International Journal of Climate Change Strategies and Management, 14*, 168.

35. Donatti, C. I., Harvey, C. A., Martinez-Rodriguez, M. R., Vignola, R., & Rodriguez, C. M. (2019). Vulnerability of smallholder farmers to climate change in Central America and Mexico: Current knowledge and research gaps. *Climate and Development, 11*(3), 264–286.
36. Dorward, P., Osbahr, H., Sutcliffe, C., & Mbeche, R. (2020). Supporting climate change adaptation using historical climate analysis. *Climate and Development, 12*(5), 469–480.
37. Dunn, M. R., Lindesay, J. A., & Howden, M. (2015). Spatial and temporal scales of future climate information for climate change adaptation in viticulture: A case study of User needs in the Australian winegrape sector. *Australian Journal of Grape and Wine Research, 21*(2), 226–239.
38. Duong, P. B., Thanh, P. T., & Ancev, T. (2021). Impacts of off-farm employment on welfare, food security and poverty: Evidence from rural Vietnam. *International Journal of Social Welfare, 30*(1), 84–96.
39. Dzanku, F. M. (2019). Food security in rural sub-Saharan Africa: Exploring the nexus between gender, geography and off-farm employment. *World Development, 113*, 26–43.
40. Eichsteller, M., Njagi, T., & Nyukuri, E. (2022). The role of agriculture in poverty escapes in Kenya–Developing a capabilities approach in the context of climate change. World Development, 149, 105705.
41. Elahi, E., Khalid, Z., Tauni, M. Z., Zhang, H., & Lirong, X. (2022). Extreme weather events risk to crop-production and the adaptation of innovative management strategies to mitigate the risk: A retrospective survey of rural Punjab, Pakistan. *Technovation, 117*, 102255.
42. FAO. (2020). *Using the food insecurity experience scale to measure the impact of COVID-19*. Report, FAO.
43. FAO, & UNICEF. (2021). *The state of food security and nutrition in the world 2021. Transforming food systems for food security, improved nutrition and affordable healthy diets for all*. FAO.
44. FAOstat, C. (2022). Livestock products. https://www.fao.org/faostat
45. Feinstein, N. W., & Mach, K. J. (2020). Three roles for education in climate change adaptation. *Climate Policy, 20*(3), 317–322.
46. Fentie, A., & Beyene, A. D. (2019). Climate-smart agricultural practices and welfare of rural smallholders in Ethiopia: Does planting method matter? *Land Use Policy, 85*, 387–396.
47. Food and Agriculture Organization of the United Nations, International Fund for Agricultural Development, United Nations Children's Fund United Nations World Food Programme, & World Health Organization. (2022). *The state of food security and nutrition in the world: 2022: Repurposing food and agricultural policies to make healthy diets more affordable*. FAO.
48. Fraval, S., Yameogo, V., Ayantunde, A., Hammond, J., De Boer, I. J., Oosting, S. J., & Van Wijk, M. T. (2020). Food security in rural Burkina Faso: The importance of consumption of own-farm sourced food versus purchased food. *Agriculture & Food Security, 9*(1), 1–17.
49. Fuglie, K., Ray, S., Baldos, U. L. C., & Hertel, T. W. (2022). The R&D cost of climate mitigation in agriculture. *Applied Economic Perspectives and Policy, 44*(4), 1955–1974.
50. Galappaththi, E. K., Ichien, S. T., Hyman, A. A., Aubrac, C. J., & Ford, J. D. (2020). Climate change adaptation in aquaculture. *Reviews in Aquaculture, 12*(4), 2160–2176.
51. Garrido, M. M., Deb, P., Burgess, J. F., Jr., & Penrod, J. D. (2012). Choosing models for health care cost analyses: Issues of nonlinearity and endogeneity. *Health Services Research, 47*(6), 2377–2397.
52. Geraci, A., Fabbri, D., & Monfardini, C. (2016). Testing exogeneity of multinomial regressors in count data models: Does two-stage residual inclusion work? *Journal of Econometric Methods, 7*(1), 20140019.
53. Gezie, M. (2019). Farmer's response to climate change and variability in Ethiopia: A review. *Cogent Food & Agriculture, 5*(1), 1613770.
54. Godde, C. M., Mason-D'Croz, D., Mayberry, D. E., Thornton, P. K., & Herrero, M. (2021). Impacts of climate change on the livestock food supply chain: A review of the evidence. *Global Food Security, 28*, 100488.
55. Griffin, C., Wreford, A., & Cradock-Henry, N. A. (2023). 'As a farmer you've just got to learn to cope': Understanding dairy farmers' perceptions of climate change and adaptation

decisions in the lower south Island of Aotearoa-New Zealand. *Journal of Rural Studies, 98*, 147–158.
56. Grossi, G., Goglio, P., Vitali, A., & Williams, A. G. (2019). Livestock and climate change: Impact of livestock on climate and mitigation strategies. *Animal Frontiers, 9*(1), 69–76.
57. Guntukula, R. (2020). Assessing the impact of climate change on Indian agriculture: Evidence from major crop yields. *Journal of Public Affairs, 20*(1), e2040.
58. Guzmán-Luna, P., Mauricio-Iglesias, M., Flysjö, A., & Hospido, A. (2022a). Analysing the interaction between the dairy sector and climate change from a life cycle perspective: A review. *Trends in Food Science & Technology, 126*, 168–179.
59. Guzmán-Luna, P., Nag, R., Martínez, I., Mauricio-Iglesias, M., Hospido, A., & Cummins, E. (2022b). Quantifying current and future raw milk losses due to bovine mastitis on European dairy farms under climate change scenarios. *Science of the Total Environment, 833*, 155149.
60. He, X., Huang, A., Yan, J., Zhou, H., Wu, Y., Yang, L. E., & Paudel, B. (2023). Smallholders' climate change adaptation strategies on the eastern Tibetan Plateau. *Natural Hazards, 118*, 1–27.
61. Headey, D. (2023). Can dairy help solve the malnutrition crisis in developing countries? An economic analysis. *Animal Frontiers, 13*(1), 7–16.
62. Hilemelekot, F., Ayal, D. Y., Ture, K., & Zeleke, T. T. (2021). Climate change and variability adaptation strategies and their implications for household food security: The case of Basona Worena District, North Shewa zone, Ethiopia. *Climate Services, 24*, 100269.
63. Horn, J., & Isselstein, J. (2022). How do we feed grazing livestock in the future? A case for knowledge-driven grazing systems. *Grass and Forage Science, 77*(3), 153–166.
64. Hossain, M. S., Arshad, M., Qian, L., Zhao, M., Mehmood, Y., & Kächele, H. (2019). Economic impact of climate change on crop farming in Bangladesh: An application of Ricardian method. *Ecological Economics, 164*, 106354.
65. Initiative, N. D. G. A. (2022). ND-GAIN Country Index. 2007.
66. Islam, M. S., Okubo, K., Islam, A. H. M. S., & Sato, M. (2022). Investigating the effect of climate change on food loss and food security in Bangladesh. *SN Business & Economics, 2*, 1–24.
67. Jamshidi, O., Asadi, A., Kalantari, K., Azadi, H., & Scheffran, J. (2019). Vulnerability to climate change of smallholder farmers in the Hamadan province, Iran. *Climate Risk Management, 23*, 146–159.
68. Jin, J., Xuhong, T., Wan, X., He, R., Kuang, F., & Ning, J. (2020). Farmers' risk aversion, loss aversion and climate change adaptation strategies in Wushen Banner, China. *Journal of Environmental Planning and Management, 63*(14), 2593–2606.
69. Joshi, D. R., Ghimire, R., Kharel, T., Mishra, U., & Clay, S. A. (2021). Conservation agriculture for food security and climate resilience in Nepal. *Agronomy Journal, 113*(6), 4484–4493.
70. Kalaugher, E., Beukes, P., Bornman, J. F., Clark, A., & Campbell, D. I. (2017). Modelling farm-level adaptation of temperate, pasture-based dairy farms to climate change. *Agricultural Systems, 153*, 53–68.
71. Kassaye, A. Y., Shao, G., Wang, X., Shifaw, E., & Wu, S. (2021). Impact of climate change on the staple food crops yield in Ethiopia: Implications for food security. *Theoretical and Applied Climatology, 145*(1–2), 327–343.
72. Kassegn, A., & Endris, E. (2021). Review on livelihood diversification and food security situations in Ethiopia. *Cogent Food & Agriculture, 7*(1), 1882135.
73. Kerr, R. B., Kangmennaang, J., Dakishoni, L., Nyantakyi-Frimpong, H., Lupafya, E., Shumba, L., Msachi, R., Boateng, G. O., Snapp, S. S., & Chitaya, A. (2019). Participatory agroecological research on climate change adaptation improves smallholder farmer household food security and dietary diversity in Malawi. *Agriculture, Ecosystems & Environment, 279*, 109–121.
74. Khanal, U., Wilson, C., Hoang, V.-N., & Lee, B. (2018). Farmers' adaptation to climate change, its determinants and impacts on rice yield in Nepal. *Ecological Economics, 144*, 139–147.

75. Khanal, U., Wilson, C., Hoang, V.-N., & Lee, B. (2019). Impact of community-based organizations on climate change adaptation in agriculture: Empirical evidence from Nepal. *Environment, Development and Sustainability, 21*, 621–635.
76. Kinda, S. R., & Badolo, F. (2019). Does rainfall variability matter for food security in developing countries? *Cogent Economics & Finance, 7*(1), 1640098.
77. Kogo, B. K., Kumar, L., & Koech, R. (2021). Climate change and variability in Kenya: A review of impacts on agriculture and food security. *Environment, Development and Sustainability, 23*(1), 23–43.
78. Kuchimanchi, B. R., van Paassen, A., & Oosting, S. J. (2021). Understanding the vulnerability, farming strategies and development pathways of smallholder farming systems in Telangana, India. *Climate Risk Management, 31*, 100275.
79. Lacetera, N. (2019). Impact of climate change on animal health and welfare. *Animal Frontiers, 9*(1), 26–31.
80. Leisner, C. P. (2020). Climate change impacts on food security-focus on perennial cropping systems and nutritional value. *Plant Science, 293*, 110412.
81. Lelea, M. A., Emden, A., & Kaufmann, B. (2023). Ensuring milk quality: Contextualizing the knowledge of Kenyan small-scale milk traders who connect farmers with low-income consumers. *Journal of Rural Studies, 100*, 103032.
82. Mabhaudhi, T., Chimonyo, V. G. P., Hlahla, S., Massawe, F., Mayes, S., Nhamo, L., & Modi, A. T. (2019). Prospects of orphan crops in climate change. *Planta, 250*, 695–708.
83. Makate, C., Wang, R., Makate, M., & Mango, N. (2016). Crop diversification and livelihoods of smallholder farmers in Zimbabwe: Adaptive management for environmental change. *SpringerPlus, 5*, 1–18.
84. Marumo, J. L., Lusseau, D., Speakman, J. R., Mackie, M., & Hambly, C. (2022). Influence of environmental factors and parity on milk yield dynamics in barn-housed dairy cattle. *Journal of Dairy Science, 105*(2), 1225–1241.
85. Masuda, Y. J., Castro, B., Aggraeni, I., Wolff, N. H., Ebi, K., Garg, T., Game, E. T., Krenz, J., & Spector, J. (2019). How are healthy, working populations affected by increasing temperatures in the tropics? Implications for climate change adaptation policies. *Global Environmental Change, 56*, 29–40.
86. Matere, J., Simpkin, P., Angerer, J., Olesambu, E., Ramasamy, S., & Fasina, F. (2020). Predictive Livestock Early Warning System (PLEWS): Monitoring forage condition and implications for animal production in Kenya. *Weather and Climate Extremes, 27*, 100209.
87. Mbuthia, J. M., Mayer, M., & Reinsch, N. (2021). Modeling heat stress effects on dairy cattle milk production in a tropical environment using test-day records and random regression models. *Animal, 15*(8), 100222.
88. Mekonnen, A., Tessema, A., Ganewo, Z., & Haile, A. (2021). Climate change impacts on household food security and farmers adaptation strategies. *Journal of Agriculture and Food Research, 6*, 100197.
89. Menghistu, H. T., Abraha, A. Z., Tesfay, G., & Mawcha, G. T. (2020). Determinant factors of climate change adaptation by pastoral/agro-pastoral communities and smallholder farmers in sub-Saharan Africa: A systematic review. *International Journal of Climate Change Strategies and Management, 12*(3), 305–321.
90. Mengistu, D. D., Degaga, D. T., & Tsehay, A. S. (2021). Analyzing the contribution of crop diversification in improving household food security among wheat dominated rural households in Sinana District, Bale Zone, Ethiopia. *Agriculture & Food Security, 10*, 1–15.
91. Mihiretu, A., Okoyo, E. N., & Lemma, T. (2019). Determinants of adaptation choices to climate change in agro-pastoral dry lands of Northeastern Amhara, Ethiopia. *Cogent Environmental Science, 5*(1), 1636548.
92. Monroe, M. C., Plate, R. R., Oxarart, A., Bowers, A., & Chaves, W. A. (2019). Identifying effective climate change education strategies: A systematic review of the research. *Environmental Education Research, 25*(6), 791–812.
93. Muchuru, S., & Nhamo, G. (2019). A review of climate change adaptation measures in the African crop sector. *Climate and Development, 11*(10), 873–885.

94. Mwinkom, F. X., Damnyag, L., Abugre, S., & Alhassan, S. I. (2021). Factors influencing climate change adaptation strategies in North-Western Ghana: Evidence of farmers in the Black Volta Basin in Upper West region. *SN Applied Sciences, 3*, 1–20.
95. Nabara, I. S., Man, N., Kamarulzaman, N. H., & Sulaiman, Z. (2021). Smallholder oil palm farmers' pro-adaptation behaviour under climate impact scenario: Application of protection Motivation Theory. *Climate and Development, 13*(6), 475–483.
96. Narita, D., Sato, I., Ogawada, D., & Matsumura, A. (2022). Evaluating the robustness of project performance under deep uncertainty of climate change: A case study of irrigation development in Kenya. *Climate Risk Management, 36*, 100426.
97. Naser, M. M., Swapan, M. S. H., Ahsan, R., Afroz, T., & Ahmed, S. (2019). Climate change, migration and human rights in Bangladesh: Perspectives on governance. *Asia Pacific Viewpoint, 60*(2), 175–190.
98. Ndayisaba, P. C., Kuyah, S., Midega, C. A. O., Mwangi, P. N., & Khan, Z. R. (2022). Push-pull technology improves carbon stocks in rainfed smallholder agriculture in Western Kenya. *Carbon Management, 13*(1), 127–141.
99. Ndiritu, S. W. (2020). Beef value chain analysis and climate change adaptation and investment options in the semi-arid lands of northern Kenya. *Journal of Arid Environments, 181*, 104216.
100. Nébié, E. K. I., Ba, D., & Giannini, A. (2021). Food security and climate shocks in Senegal: Who and where are the most vulnerable households? *Global Food Security, 29*, 100513.
101. Niles, M. T., Ahuja, R., Barker, T., Esquivel, J., Gutterman, S., Heller, M. C., Mango, N., Portner, D., Raimond, R., & Tirado, C. (2018). Climate change mitigation beyond agriculture: A review of food system opportunities and implications. *Renewable Agriculture and Food Systems, 33*(3), 297–308.
102. Nthambi, M., & Ijioma, U. D. (2021). Retracing economic impact of climate change disasters in Africa: Case study of drought episodes and adaptation in Kenya. In *African handbook of climate change adaptation* (pp. 1007–1031). Springer.
103. Nyariki, D. M., & Amwata, D. A. (2019). The value of pastoralism in Kenya: Application of total economic value approach. *Pastoralism, 9*(1), 1–13.
104. Nyiwul, L. (2021). Climate change adaptation and inequality in Africa: Case of water, energy and food insecurity. *Journal of Cleaner Production, 278*, 123393.
105. Ojo, T. O., & Baiyegunhi, L. J. S. (2020). Determinants of credit constraints and its impact on the adoption of climate change adaptation strategies among rice farmers in South-West Nigeria. *Journal of Economic Structures, 9*(1), 1–15.
106. Oluoch, K. O., De Groote, H., Gitonga, Z. M., Jin, Z., & Davis, K. F. (2022). A suite of agronomic factors can offset the effects of climate variability on rainfed maize production in Kenya. *Scientific Reports, 12*(1), 16043.
107. Omerkhil, N., Chand, T., Valente, D., Alatalo, J. M., & Pandey, R. (2020). Climate change vulnerability and adaptation strategies for smallholder farmers in Yangi Qala District, Takhar, Afghanistan. *Ecological Indicators, 110*, 105863.
108. Osiemo, J., Girvetz, E. H., Hasiner, E., Schroeder, K., Treguer, D., Juergenliemk, A., Horst, A., Jarvis, A., & Kropff, W. (2021). Digital agriculture profile: Kenya.
109. Owusu, V., Ma, W., Renwick, A., & Emuah, D. (2021). Does the use of climate information contribute to climate change adaptation? Evidence from Ghana. *Climate and Development, 13*(7), 616–629.
110. Panda, A. (2016). Exploring climate change perceptions, rainfall trends and perceived barriers to adaptation in a drought affected region in India. *Natural Hazards, 84*, 777–796.
111. Partey, S. T., Dakorah, A. D., Zougmoré, R. B., Ouédraogo, M., Nyasimi, M., Nikoi, G. K., & Huyer, S. (2020). Gender and climate risk management: Evidence of climate information use in Ghana. *Climatic Change, 158*, 61–75.
112. Programme, W. F. (2015). *Food consumption score nutritional quality analysis guidelines (FCS-N)*. United Nations World Food Programme, Food Security Analysis (VAM).
113. Rahman, M. S., Huang, W.-C., Toiba, H., & Efani, A. (2022a). Does adaptation to climate change promote household food security? Insights from Indonesian fishermen. *International Journal of Sustainable Development & World Ecology, 29*(7), 611–624.

114. Rahman, M. S., Zulfiqar, F., Ullah, H., Himanshu, S. K., & Datta, A. (2022b). Status and drivers of households' food security status in climate-sensitive coastal areas of Bangladesh: A comparison between the exposed and interior coasts. *International Journal of Sustainable Development & World Ecology*, 1–14.
115. Ranasinghe, R. D. A. K., Korale-Gedara, P. M., & Weerasooriya, S. A. (2023). Climate change adaptation and adaptive capacities of dairy farmers: Evidence from village tank cascade systems in Sri Lanka. *Agricultural Systems, 206*, 103609.
116. Reppin, S., Kuyah, S., de Neergaard, A., Oelofse, M., & Rosenstock, T. S. (2020). Contribution of agroforestry to climate change mitigation and livelihoods in Western Kenya. *Agroforestry Systems, 94*(1), 203–220.
117. Savari, M., & Zhoolideh, M. (2021). The role of climate change adaptation of small-scale farmers on the households food security level in the west of Iran. *Development in Practice, 31*(5), 650–664.
118. Schneider, P., & Asch, F. (2020). Rice production and food security in Asian Mega deltas—A review on characteristics, vulnerabilities and agricultural adaptation options to cope with climate change. *Journal of Agronomy and Crop Science, 206*(4), 491–503.
119. Scoones, I. (2023). Livestock, methane, and climate change: The politics of global assessments. *Wiley Interdisciplinary Reviews: Climate Change, 14*(1), e790.
120. Sertse, S. F., Khan, N. A., Shah, A. A., Liu, Y., & Naqvi, S. A. A. (2021). Farm households' perceptions and adaptation strategies to climate change risks and their determinants: Evidence from Raya Azebo district, Ethiopia. *International Journal of Disaster Risk Reduction, 60*, 102255.
121. Singh, S. (2020). Farmers' perception of climate change and adaptation decisions: A micro-level evidence from Bundelkhand Region, India. *Ecological Indicators, 116*, 106475.
122. Sridhar, A., Balakrishnan, A., Jacob, M. M., Sillanpää, M., & Dayanandan, N. (2023). Global impact of COVID-19 on agriculture: Role of sustainable agriculture and digital farming. *Environmental Science and Pollution Research, 30*(15), 42509–42525.
123. Srivastava, Y. (2019). Climate change: A challenge for postharvest management, food loss, food quality, and food security. In *Climate Change and agricultural ecosystems* (pp. 355–377). Elsevier.
124. Suman, M. (2022). Post-disaster interventions related to livestock and livelihood. In *Management of Animals in disasters* (pp. 87–98). Springer.
125. Sutcliffe, C., Dougill, A. J., & Quinn, C. H. (2016). Evidence and perceptions of rainfall change in Malawi: Do maize cultivar choices enhance climate change adaptation in sub-Saharan Africa? *Regional Environmental Change, 16*, 1215–1224.
126. Terza, J. V. (2018). Two-stage residual inclusion estimation in health services research and health economics. *Health Services Research, 53*(3), 1890–1899.
127. Thornton, P., Nelson, G., Mayberry, D., & Herrero, M. (2022). Impacts of heat stress on global cattle production during the 21st century: A modelling study. *The Lancet Planetary Health, 6*(3), e192–e201.
128. Tofu, D. A., Woldeamanuel, T., & Haile, F. (2022). Smallholder farmers' vulnerability and adaptation to climate change induced shocks: The case of Northern Ethiopia highlands. *Journal of Agriculture and Food Research, 8*, 100312.
129. Van Der Pol, T. D., Van Ierland, E. C., Gabbert, S., Weikard, H.-P., & Hendrix, E. M. T. (2015). Impacts of rainfall variability and expected rainfall changes on cost-effective adaptation of water systems to climate change. *Journal of Environmental Management, 154*, 40–47.
130. Walker, S. E., Bruyere, B. L., Zarestky, J., Yasin, A., Lenaiyasa, E., Lolemu, A., & Pickering, T. (2022). Education and adaptive capacity: The influence of formal education on climate change adaptation of pastoral women. *Climate and Development, 14*(5), 409–418.
131. Wamsler, C., Brink, E., & Rentala, O. (2012). Climate change, adaptation, and formal education: The role of schooling for increasing societies' adaptive capacities in El Salvador and Brazil. *Ecology and Society, 17*(2).
132. Wan, F., Small, D., & Mitra, N. (2018). A general approach to evaluating the bias of 2-stage instrumental variable estimators. *Statistics in Medicine, 37*(12), 1997–2015.

133. Wichern, J., Descheemaeker, K., Giller, K. E., Ebanyat, P., Taulya, G., & van Wijk, M. T. (2019). Vulnerability and adaptation options to climate change for rural livelihoods–A country-wide analysis for Uganda. *Agricultural Systems, 176*, 102663.
134. Wiebe, K., Robinson, S., & Cattaneo, A. (2019). Climate change, agriculture and food security: Impacts and the potential for adaptation and mitigation. *Sustainable Food and Agriculture*, 55–74.
135. Winning, M., Price, J., Ekins, P., Pye, S., Glynn, J., Watson, J., & McGlade, C. (2019). Nationally determined contributions under the Paris agreement and the costs of delayed action. *Climate Policy, 19*(8), 947–958.
136. Yan, S., & Alvi, S. (2022). Food security in South Asia under climate change and economic policies. *International Journal of Climate Change Strategies and Management, 14*(3), 237–251.
137. Ying, A., Xu, R., & Murphy, J. (2019). Two-stage residual inclusion for survival data and competing risks—An instrumental variable approach with application to SEER-Medicare linked data. *Statistics in Medicine, 38*(10), 1775–1801.
138. Zahid, M., Rahman, H. U., Khan, M., Ali, W., & Shad, F. (2020). Addressing endogeneity by proposing novel instrumental variables in the nexus of sustainability reporting and firm financial performance: A step-by-step procedure for non-experts. *Business Strategy and the Environment, 29*(8), 3086–3103.
139. Zamasiya, B., Nyikahadzoi, K., & Mukamuri, B. B. (2017). Factors influencing smallholder farmers' behavioural intention towards adaptation to climate change in transitional climatic zones: A case study of Hwedza District in Zimbabwe. *Journal of Environmental Management, 198*, 233–239.
140. Zavala Nacul, H., & Revoredo-Giha, C. (2022). Food safety and the informal milk supply chain in Kenya. *Agriculture & Food Security, 11*(1), 1–14.
141. Zegeye, M. B., & Meshesha, G. B. (2022). Estimating the impact of fertilizer adoption on poverty in rural Ethiopia: An endogenous switching regression approach. *Asia-Pacific Journal of Regional Science, 6*, 1–21.
142. Zhu, Z., Ma, W., & Leng, C. (2020). ICT adoption, individual income and psychological health of rural farmers in China. *Applied Research in Quality of Life*, 1–21.

Resilience and Adaptation Strategies Among Women in Response to Climate Change in Wajir West Sub-County, Kenya

Nicolette Vanessa Roman, Moses Mutua Mutiso, Hamdi Abdi, and Claire Mutsilila Nakweya

Abstract

Climate change poses significant challenges to communities in Wajir West sub-county, Kenya, with women disproportionately affected due to their caregiving, livelihood, and household management roles. This study examines women's resilience and adaptation strategies in response to recurrent droughts, floods, and resource scarcity, guided by the Sustainable Livelihoods Framework (SLF). A qualitative research design was adopted, involving in-depth interviews with 16 women and 2 focus group discussions (FGDs). Data were thematically analyzed using NVivo software, ensuring confidentiality and informed consent throughout. The findings reveal five key adaptation themes. Economic adaptation strategies include livelihood diversification, where women engage in small-scale trade, firewood collection, and informal market sales. Community-based support networks such as savings groups (chamas) provide emergency loans for healthcare, education, and food security. Psychological resilience is sustained through peer support, religious faith, and psychosocial support programs. Disaster risk reduction strategies involve flood-proofing homes, early evacuation, and collec-

Authors Nicolette Vanessa Roman, Moses Mutua Mutiso, Hamdi Abdi and Claire Mutsilila Nakweya contributed equally to this work.

N. V. Roman
Centre for Interdisciplinary Studies of Children, Families and Society (CISCFS), Faculty of Community and Health Sciences, University of the Western Cape, Cape Town, South Africa
e-mail: nroman@uwc.ac.za

M. M. Mutiso (✉) · H. Abdi · C. M. Nakweya
Department of Sociology, Psychology and Anthropology, Moi University, Eldoret, Kenya
e-mail: mosesmutua@mu.ac.ke

N. Ronoh, C. Rono (eds.), *Rethinking Our Ecosystems*, Springer Proceedings in Geography and Sustainability, https://doi.org/10.1007/978-3-031-95781-9_10

tive resource pooling. Health and nutrition strategies include reliance on herbal remedies, dietary diversification, and healthcare financing through savings groups. Despite these efforts, women face barriers such as limited financial access, gender-based constraints, and physical and emotional strain. The study underscores the need for policies that enhance women's financial inclusion, promote livelihood diversification, and strengthen gender-responsive climate adaptation interventions.

Keywords

Climate change · Women's resilience · Adaptation strategies · Wajir West · Kenya · Gender-sensitive policies

1 Introduction

Climate change is one of the most pressing global challenges of the twenty-first century, posing significant socio-economic, environmental, and health risks [1, 2]. The increase in extreme weather events such as floods, droughts, and storms has far-reaching effects on livelihoods, ecosystems, and human well-being. Rising global temperatures, shifting precipitation patterns, and increasing sea levels have escalated the frequency and severity of natural disasters. The United Nations Framework Convention on Climate Change (UNFCCC) highlights that the adverse effects of climate change are disproportionately experienced by vulnerable groups, particularly women, children, and marginalized communities [3, 4].

Women, especially in developing countries, bear the brunt of climate change impacts due to their roles as caregivers, food producers, and water collectors [5–7]. These roles make them more vulnerable to climate-induced disruptions, particularly in arid and semiarid areas. While men often migrate to urban centers or other areas to seek alternative livelihoods, women remain at home to manage household affairs, making them more exposed to climate-related stressors. Given the growing recognition of gender-specific vulnerabilities, the role of women in climate change adaptation and resilience-building has become a focal point of international climate policies and development agendas [8].

Sub-Saharan Africa is one of the regions most vulnerable to climate change, with arid and semiarid areas experiencing the greatest impacts [9–11]. The Intergovernmental Panel on Climate Change (IPCC) reports that Africa's temperature has been rising faster than the global average, with extreme droughts and floods becoming more frequent. These climatic changes have disrupted agricultural production, water availability, and economic livelihoods, particularly in pastoralist and agropastoralist communities [12]. In pastoral communities, such as those found in Kenya's arid and semiarid lands (ASALs), climate change has triggered prolonged droughts that have decimated livestock populations the primary source of wealth, nutrition, and income for households [13–15]. At the same time, flash floods have caused widespread destruction of homes, displacement of families, and loss of agricultural assets. As the backbone of many rural economies, women are disproportionately affected by these climate shocks, as they are responsible for securing water, food, and fuelwood for their households. Studies show that women's workloads increase significantly during climate-induced crises, exposing them to physical exhaustion and emotional strain [16, 17].

Kenya's ASALs, which cover approximately 89% of the country's landmass, are highly susceptible to climate-induced disasters [18–20]. Wajir county, located in northeastern Kenya, exemplifies the unique vulnerabilities faced by ASAL communities. The region is predominantly inhabited by pastoralist communities that rely on livestock for their livelihood. However, recurrent droughts have depleted pastures and water sources, leading to widespread loss of livestock. Conversely, floods periodically sweep through the region, causing displacement, destruction of homes, and disruption of livelihoods [21–24]. The socio-economic impacts of these climate-induced shocks are particularly severe for women, as they are tasked with managing household well-being in the absence of male family members, who often migrate in search of employment or greener pastures for livestock [16, 25]. Women are left to provide food, water, and healthcare for children, the elderly, and other vulnerable household members. The heightened responsibilities expose women to emotional stress, financial pressure, and health risks [5]. Furthermore, limited access to financial resources, formal education, and decision-making power constrains women's capacity to adapt effectively to climate shocks.

The role of women in climate adaptation and resilience has become a central theme in global climate policies, including the United Nations' Sustainable Development Goals (SDGs) [8, 26]. The SDG 13 on climate action and SDG 5 on gender equality emphasize the need to strengthen women's capacity to adapt to climate change [27, 28]. Women's unique knowledge of household resource management, caregiving, and food security positions them as critical actors in climate resilience efforts [26, 29]. Research highlights that women's capacity to adapt is enhanced by access to social capital, financial services, and livelihood diversification strategies [30, 31]. Women-led savings groups have emerged as a key mechanism for building financial resilience, enabling women to access loans for healthcare, school fees, and livelihood investments. Additionally, women's reliance on traditional knowledge systems such as herbal medicine and indigenous food systems has been critical in ensuring health and nutrition during climate shocks [30–32].

While existing research underscores the importance of gender-responsive climate adaptation, there remain gaps in understanding the specific adaptation and resilience strategies employed by women in ASAL regions like Wajir West. Much of the existing literature focuses on macro-level climate policies and formal interventions, with limited attention paid to the everyday, community-driven actions of women [19, 33, 34]. Women's adaptation strategies, such as reliance on savings groups, informal labor, and community-based support networks, are underexplored.

This study aimed to address this gap by investigating the adaptation and resilience strategies adopted by women in Wajir West. By focusing on the lived experiences and practical responses of women, the study sought to generate context-specific insights that can inform gender-sensitive climate adaptation strategies. The study will contribute to existing knowledge by highlighting community-driven actions and informal coping mechanisms often overlooked in policy discourses. By centering women's experiences, the study aimed to provide a holistic understanding of how local adaptation strategies intersect with broader climate resilience goals.

2 Theoretical Framework

The theoretical framework for this study on women's adaptation and resilience strategies in response to climate change in Wajir West sub-county was based on the Sustainable Livelihoods Framework (SLF) developed by Chambers and Conway [35]. The SLF provides a comprehensive approach for understanding how communities mobilize diverse resources to sustain livelihoods amid external shocks such as climate change. Central to this framework are five key livelihood capitals: human, natural, physical, financial, and social. Human capital includes skills, knowledge, and labor, which women in Wajir West use in caregiving, small-scale trade, and indigenous health support. Natural capital refers to access to land, water, and forests, which women utilize for collecting firewood, growing indigenous crops, and harvesting wild fruits. Physical capital consists of tangible resources like flood-proof homes, water storage tanks, and cooking equipment, which enhance household resilience. Financial capital includes savings, credit, and access to financial support, often facilitated through savings groups (chamas) that provide women with loans for education, healthcare, and emergencies. Lastly, social capital is represented by support networks and community relationships that women rely on to access resources and assistance during crises [35, 36].

The SLF highlights how women in Wajir West use various livelihood capitals to strengthen resilience and adapt to climate change. The role of financial and social capital is seen in savings groups, which provide financial support for essential needs like healthcare and education, while also fostering social cohesion and collective action [35]. Human capital is demonstrated by women's shift from pastoral livelihoods to small-scale trade, reflecting their capacity to acquire new skills necessary for livelihood diversification [36]. Natural capital is evidenced in women's use of forests for firewood and wild fruits, which sustain household food security during periods of scarcity. These adaptation practices align with findings from other regions, such as Bangladesh and Ghana, where women use natural capital to sustain food security during climate shocks [37–39].

The SLF has several advantages in understanding climate adaptation. Its holistic approach captures the dynamic interactions between livelihood capitals, offering a more comprehensive analysis of women's adaptation strategies. While traditional economic models emphasize income alone, the SLF incorporates social, physical, and natural dimensions. For instance, women's ability to draw on multiple capitals, such as social (savings groups) and natural (wild fruits), demonstrates how multiple resources interact to strengthen resilience [36, 40–43]. The SLF also recognizes that livelihoods are dynamic and shift in response to external shocks. This dynamic perspective is relevant in Wajir West, where women have transitioned from livestock production to informal trade due to recurrent droughts. Additionally, the SLF highlights women's agency, portraying them as active decision-makers who adapt, cope, and recover from climate challenges [28, 35, 44]

However, the SLF is not without limitations. It does not explicitly address how power relations and gender inequalities shape access to livelihood resources. In Wajir West, women have limited control over natural capital, as land and natural resources are often controlled by men. This limits women's capacity to fully utilize natural capital for farming or other livelihood activities. Additionally, financial capital is not accessible to all women. While savings groups offer support, only group members benefit, leaving nonmembers with limited financial options [32, 45–47]. Another limitation is the SLF's tendency to view livelihood capitals as independent entities, whereas, in practice, they are interdependent. For example, physical capital (like water tanks) is linked to natural capital (like water access) and financial capital (like the cost of water purchases). This lack of focus on interdependencies limits the framework's ability to capture the complexities of women's adaptation strategies [39].

Despite these limitations, the SLF remains a valuable tool for examining women's adaptation and resilience strategies. It offers an in-depth understanding of how women use human capital (herbal medicine) and financial capital (savings group loans) to address health shocks caused by climate change. It also explains how women combine natural capital (wild fruits) and human capital (health knowledge) to create alternative health solutions when formal healthcare is inaccessible. The SLF's emphasis on dynamic livelihood transitions is relevant to women's shift from livestock production to informal trade, a trend also observed in Ethiopia, Bangladesh, and Ghana [48–50]. This approach enables researchers and policymakers to analyze women's adaptation as a continuous process of change, capturing the transformative nature of resilience in the face of climate challenges.

The SLF also offers insights for policy and practice. Policymakers can strengthen women's access to livelihood capitals by promoting financial inclusion, supporting savings groups, and increasing access to affordable credit and microfinance. Additionally, enhancing women's access to physical capital, such as flood-proof housing and water storage facilities, could reduce post-disaster recovery costs and support household resilience. Programs promoting climate-resilient agriculture, such as drought-resistant crops, could strengthen access to natural capital [36]. Training programs aimed at improving human capital, such as disaster preparedness, mental health support, and entrepreneurial skills, would enable women to better manage the demands of caregiving, business, and emergency response. Development actors can also foster social capital by supporting women's groups and cooperative networks, enabling collective action in times of crisis [32].

In summary, the Sustainable Livelihoods Framework (SLF) provided a comprehensive approach for understanding how women in Wajir West adapt to and build resilience against climate change. It highlights the role of human, natural, physical, financial, and social capital in women's efforts to strengthen household resilience. The SLF captures the dynamic nature of livelihoods, showing how women shift from pastoralism to informal trade, health provision, and diversified income-generating activities in response to climate shocks. Despite its limitations, such as its insufficient focus on gender inequalities, the SLF remains a valuable lens for analyzing gendered adaptation processes. By supporting financial inclusion, enhancing access to livelihood capitals, and fostering social networks, policymakers and development actors can promote sustainable and gender-sensitive climate adaptation strategies. The SLF's emphasis on livelihood resilience offers a practical framework for designing interventions that empower women and promote sustainable development in climate-vulnerable regions like Wajir West [32, 35, 36].

3 Methodology

This study investigated the adaptation and resilience strategies employed by women in Wajir West sub-county, Kenya, in response to climate change. The choice of Wajir West as the study site is informed by its high vulnerability to climate-induced shocks, such as droughts and floods, which severely disrupt livelihoods and family structures. Women were chosen as the target population due to their critical roles in household resource management and their significant contribution to climate adaptation at the household and community levels. The study sample consisted of 16 women purposefully selected to ensure diverse perspectives from participants who had direct experiences with climate-induced challenges and were actively engaged in adaptation and resilience-building efforts.

Data collection utilized a combination of in-depth interviews and focus group discussions (FGDs) to capture both individual and collective perspectives. Sixteen women participated in one-on-one in-depth interviews, allowing for a detailed exploration of their unique experiences, challenges, and strategies for adaptation. To complement the individual insights, two FGDs were conducted, each comprising eight participants from diverse socio-economic, age, and marital status backgrounds. The diverse composition of the groups enabled the study to gather a wide range of experiences and foster collective reflection on community-driven responses to climate shocks.

Data analysis followed a thematic analysis approach, a method particularly suitable for qualitative data. This approach facilitated the identification, categorization, and interpretation of key themes related to adaptation strategies, livelihood diversification, community-based support systems, and family resilience. The audio recordings of interviews and FGDs were transcribed verbatim to ensure accuracy. A coding framework was developed to categorize the data into core themes and sub-themes, with NVivo software used to organize, code, and analyze the data systematically. This approach enabled the identification of recurring themes and crosscutting issues that reflect the lived experiences of women in Wajir West.

Ethical considerations were a core part of the research process, ensuring the dignity, privacy, and well-being of participants. Each participant was fully briefed on the objectives, procedures, and ethical principles guiding the study. Informed consent was obtained before participation, with assurances that respondents had the right to withdraw at any point without consequences. To protect participant privacy, personal identifiers were anonymized, and all data were securely stored. Furthermore, the study adhered to confidentiality protocols, ensuring that participants' information was accessible only to the research team. Ethical approval for the study was granted by the National Commission for Science, Technology, and Innovation (NACOSTI), providing further assurance of compliance with research ethics and regulatory standards.

This methodological approach was designed to generate a comprehensive understanding of women's adaptation and resilience strategies to climate change. The use of in-depth interviews and focus group discussions facilitated the collection of rich, contextualized data. The ethical safeguards ensured participant protection, while the use of thematic analysis and NVivo software enabled a systematic and transparent approach to data management and analysis. This approach provided a nuanced understanding of the lived experiences of women in Wajir West, contributing to knowledge on gender-responsive climate adaptation and resilience.

4 Results and Discussion

The demographic profile of respondents in Wajir West sub-county provided crucial context for understanding their adaptation and resilience strategies to climate change. The age distribution showed that 50% of respondents were aged 21 to 42, while 37.5% were aged 43 to 63, with only 6.2% under 20 or over 64. This indicated that most respondents were in their prime caregiving and working years, a period often associated with heightened economic and household responsibilities. Marital status data revealed that 37.5% of respondents were married, an equal percentage were single, 18.8% were widowed, and 6.2% were divorced. Widowed and divorced women, in particular, often faced higher caregiving and financial burdens due to the absence of male support within the household.

Household structure data indicated that 56.2% of respondents lived in polygamous households, while 43.8% lived in nuclear families. Polygamous households often faced greater strain as limited resources had to be shared among multiple wives and children. Family size added further complexity, with 68.8% of respondents having children. Of these, 31.2% had between five and eight children, and 12.5% had nine or more. Households with large family sizes were more susceptible to food insecurity and economic strain, especially during climate-induced shocks like droughts and floods. All respondents identified as Somali and Muslim, reflecting strong adherence to cultural and religious norms, which often influence gender roles and family responsibilities.

Educational attainment was a key determinant of women's ability to adapt to shifting gender roles and climate shocks. Among respondents, 25% had no formal education, 18.8% had completed primary education, 37.5% had secondary education, and 18.8% had tertiary qualifications. Limited education restricted access to alternative livelihoods and formal employment, leaving 31.2% of respondents unemployed and 31.2% as housewives. Only 37.5% were self-employed, often in informal economic activities like small-scale trading and caregiving roles. This limited access to formal employment increased women's vulnerability to economic shocks, especially as climate change disrupted traditional livelihood sources.

These demographic characteristics provided a foundation for analyzing the study's core findings, which are presented in Table 1. The findings highlight five key themes: economic adaptation strategies, community-based support networks, psychological resilience and emotional well-being, disaster risk reduction and emergency response strategies, and health and nutrition strategies. These themes illustrate how women in Wajir West sub-county employ diverse strategies to strengthen household resilience, manage caregiving roles, and sustain livelihoods amid climate-induced disruptions. The findings underscore the central role of women as key agents of adaptation, navigating the dual pressures of shifting gender roles and climate-related challenges.

4.1 Theme 1: Economic Adaptation Strategies

Climate change in Wajir West has significantly disrupted traditional livelihoods, particularly pastoralism, due to recurrent droughts and floods that decimate livestock, a vital source of household wealth and nutrition. To cope with these shocks, women have diversified their income sources, marking a shift in their economic roles. Many have engaged in small-scale trading, selling firewood, milk, and vegetables at local markets. While this shift highlights their adaptability, it also exposes them to economic instability and physical strain. Unlike formal employment, informal trade is unpredictable, physically demanding, and often insecure, forcing women to balance caregiving with income generation.

> "After the drought killed most of our livestock, I started selling firewood. It's hard because sometimes we have to walk for hours to find firewood, but at least I can sell some and buy food for my children." (Participant 4, individual interview)

This testimony reveals the physical burden of firewood collection, which also exposes women to risks like gender-based violence. Despite these dangers, women persist to ensure household survival.

> "I now sell milk and vegetables at the local market. It's not what I used to do, but I had to start because my husband is away most of the time. The money is small, but it helps us get by." (Participant 2, focus group discussion)

This narrative highlight how men's migration forces women to become breadwinners, leading to shifts in household power dynamics. These changes can create conflict when men return and attempt to reclaim decision-making roles.

> "There are days when I have nothing to sell, and that means my children don't eat. This business is unstable, but it is the only option I have right now." (Participant 8, individual interview)

Table 1 Resilience and adaptation strategies among women in response to climate change

Theme	Sub-themes
Economic adaptation strategies	Livelihood diversification (e.g., small-scale trade, selling firewood, vegetables, and milk)
	Informal trade and income generation
	Shift from pastoralism to alternative livelihoods
	Women's role as breadwinners and economic providers
Community-based support networks	Participation in savings groups (chamas)
	Collective resource pooling
	Access to emergency loans for education, healthcare, and shelter
	Strengthening social networks and peer support
Psychological resilience and emotional well-being	Peer-based emotional support (e.g., social support from savings groups)
	Faith and religious coping (e.g., prayer and spiritual reflection)
	Routine planning and task prioritization
	Self-care and leisure activities (e.g., playing with children, and attending social events)
	Community-based mental health and psychosocial support (e.g., MHPSS training)
Disaster risk reduction and emergency response strategies	Household risk preparedness (e.g., emergency kits and pre-positioning food and water)
	Early evacuation and relocation planning (e.g., relocation to higher ground)
	Flood-proofing homes and assets (e.g., sealing cracks and raising furniture)
	Community-based preparedness systems (e.g., collective planning and shared resources)
	Post-disaster recovery and rebuilding (e.g., cleaning homes and repairing roofs)
Health and nutrition strategies	Use of traditional herbal remedies (e.g., treatment of diarrhea and skin diseases)
	Nutritional diversification (e.g., wild fruits, indigenous crops, and drought-resistant crops)
	Healthcare financing through savings groups (e.g., emergency loans for medical expenses)
	Psychosocial support (e.g., support from women's groups and religious support)
	Community-based health interventions (e.g., use of herbal knowledge for healthcare)

This statement underscores the precarious nature of informal trade. With no guaranteed income, women are vulnerable to supply disruptions, market fluctuations, and reduced purchasing power, leaving them unable to save or plan for future shocks.

The findings reveal key insights into women's economic adaptation. Their engagement in informal trade demonstrates resilience and adaptability, but it also exposes them to financial insecurity. The dual role of caregiving and breadwinning increases their physical and emotional burden, while changes in gender roles disrupt household power dynamics. These shifts challenge traditional norms, often leading to conflict when men return from migration. These findings align with studies from other regions. Addo et al. [51] and Ndumbe [52] found that women in West Africa engaged in informal trade, such as selling firewood and vegetables, as an adaptation strategy. Similar findings were reported by Chowdhury et al. [53] in Bangladesh, where women displaced by floods turned to petty trade but faced fluctuating market prices, low returns, and limited financial support. In Ethiopia, Haile [54] and Scheurlen [55] highlighted women's reliance on firewood collection as an income source after droughts, with physical strain and exposure to violence as key risks. The cumulative evidence underscores the global nature of women's adaptation strategies through livelihood diversification in response to climate shocks. The policy implications are significant. Development partners and policymakers should prioritize alternative livelihood support, such as capacity-building, microfinance, and small business development. Providing women with access to clean energy technologies, like fuel-efficient stoves, could reduce their reliance on firewood collection, addressing safety risks and environmental concerns. Formal financial support and training in sustainable livelihoods would enable women to shift from precarious informal trade to more stable income-generating activities, strengthening household resilience to future climate-induced shocks.

4.2 Theme 2: Community-Based Support Networks

Climate change in Wajir West has intensified pressures on households, particularly for women, who are responsible for household management and income generation. Recurrent droughts and floods have led to livestock losses, reduced agricultural activities, and limited access to water and firewood. These shocks leave women vulnerable to food insecurity, healthcare emergencies, and the inability to pay school fees. To cope, women have increasingly relied on community-based support networks, especially savings groups (locally known as *chamas*). These informal cooperatives allow women to save money collectively and access loans during emergencies. Unlike formal financial institutions, savings groups have low-entry barriers, operate on trust, and offer immediate access to funds, enabling women to adapt to climate shocks and build household resilience.

> "When the drought came, most of our livestock died, and I didn't know how I would pay for my children's school fees. But through our savings group, I borrowed money, and I was

> able to pay for my daughter's fees on time. Without it, she would have missed school." (Participant 3, individual interview)

This account highlights how climate-induced droughts disrupt household incomes, especially for families dependent on pastoralism. By accessing loans from savings groups, women ensure their children continue their education, preserving long-term household stability and resilience.

> "During the floods, our house was damaged, and we had no place to stay. I asked for support from my savings group, and they loaned me money to repair the house. I don't know how I would have managed without it." (Participant 7, focus group discussion)

This narrative illustrates how climate-induced floods damage housing and disrupt living conditions. Women access savings group loans to repair their homes, ensuring family safety and stability.

> "When my child became sick after the drought, I had no money to take him to the clinic. I reached out to my savings group, and they gave me money to pay for the treatment. If I had to borrow from the local shop, I would be paying back with interest, but in the group, I didn't have to pay interest." (Participant 5, individual interview)

This statement demonstrates how savings groups support women's access to healthcare. Without this support, women would be forced to rely on high-interest loans from informal lenders, further deepening household debt.

The findings reveal that savings groups act as rapid financial response systems, financial buffers for livelihood loss, and platforms for long-term financial planning. These groups allow women to access emergency loans, reducing dependence on exploitative credit options. During floods, women quickly repair homes using savings group loans, thereby minimizing disruption. In times of livelihood loss, such as the death of livestock, savings groups provide financial support, enabling women to sustain household incomes while exploring alternative livelihoods. Women also use savings groups for proactive planning by contributing regularly to create a financial cushion for future climate shocks. This foresight prevents them from being caught off guard, allowing them to better manage health, education, and livelihood needs. Global evidence aligns with the experiences of women in Wajir West. Zoynul and Fahmida [56] found that women in Bangladesh accessed credit from savings groups to fund education and housing repairs following floods. Berg [57] noted that women-led savings groups in Southeast Asia played a pivotal role in household recovery after floods and cyclones. Lukwa et al. [58] observed that Ethiopian women used emergency loans from savings groups to cover healthcare expenses and rebuild their livelihoods after droughts. The experiences of women in Wajir West mirror these findings, illustrating the global role of savings groups as key resilience mechanisms in climate-vulnerable areas. The policy implications are significant. Policymakers should support savings groups as critical tools for climate adaptation. Strengthening these groups through technical assistance, capacity-building, and access to microfinance will enhance women's financial resilience. Formalizing savings groups as cooperatives would enable access to larger loans from development partners. Integrating savings groups into climate adaptation strategies would ensure timely access to financial resources, reducing reliance on high-interest loans and strengthening household recovery. By supporting women-led savings groups, policymakers can create sustainable, gender-sensitive financial inclusion strategies for climate adaptation.

4.3 Theme 3: Psychological Resilience and Emotional Well-Being

Climate change in Wajir West has intensified the psychological burden on women, affecting their emotional well-being and resilience. Women face the "triple burden" of managing caregiving, household chores, and income generation as men migrate for work. This increased workload exposes them to emotional distress, anxiety, and mental exhaustion. Despite these challenges, women in Wajir West have demonstrated psychological resilience by adopting multiple coping strategies to sustain their emotional well-being and household stability.

One key strategy is peer-based emotional support. Women rely on informal social networks to share struggles, offer advice, and provide emotional encouragement. Participation in women-led savings groups provides not only financial support but also emotional solidarity.

> "Whenever I feel like I am about to give up, I meet with other women in the savings group. We talk about our problems, and I realize I am not alone. It gives me hope to keep going." (Participant 5, focus group discussion)

These peer networks serve as informal counseling spaces, reducing isolation, promoting mental well-being, and strengthening women's capacity to manage daily challenges. This is consistent with findings in Ghana [59] and Senegal [60], where women's participation in peer networks reduced emotional strain and anxiety linked to climate shocks.

Religious and spiritual practices also play a key role in promoting emotional resilience. Women rely on prayer, religious gatherings, and spiritual reflection to find emotional solace and acceptance of challenges beyond their control.

> "When I feel like I have reached my limit, I pray. It gives me peace of mind and the strength to keep going. Without prayer, I would have given up long ago." (Participant 2, individual interview)

This practice mirrors trends in Bangladesh, where displaced women engaged in collective prayer to manage the emotional strain of displacement [61]. By drawing strength from faith, women in Wajir West reduce anxiety and maintain emotional stability amid caregiving, livelihood stress, and unpredictable climate shocks.

Another important strategy is routine planning and prioritization. Women structure their daily tasks to reduce mental strain and create time for rest. This approach enables them to anticipate and manage daily responsibilities while ensuring flexibility in case of emergencies.

> "I make sure I fetch water early in the morning and prepare food before noon. That way, I have time in the afternoon to rest or look for extra work if needed." (Participant 8, individual interview)

By planning ahead, women reduce emotional fatigue and feel more in control of their responsibilities. This strategy aligns with findings from Senegal [62] and Ghana [63], where routine planning and daily task management were found to reduce emotional stress and improve women's mental well-being.

Self-care practices are also vital for managing emotional strain. Women create opportunities for leisure and relaxation, such as spending time with children, listening to music, or attending social gatherings like weddings or religious celebrations.

> "Sometimes I take my children to the river to wash clothes, and while we are there, we play together. It helps me forget my worries, even if it's just for a short time." (Participant 4, individual interview)

These moments of leisure provide emotional relief from daily stressors, reduce emotional fatigue, and strengthen family bonds. Similar practices have been observed in Bangladesh, where communal cooking and social activities among women were found to strengthen emotional resilience [64].

Support from community-based organizations (CBOs) and nongovernmental organizations (NGOs) further strengthens women's emotional resilience. These organizations provide training on mental health and psychosocial support (MHPSS) as part of broader climate adaptation programs. Through these programs, women learn practical stress management techniques and emotional regulation strategies.

> "We attended a training on stress management, and I learned breathing exercises that help me calm down when I feel too stressed. It works, especially when my children are crying and I don't know what to do." (Participant 7, focus group discussion)

These capacity-building initiatives enable women to manage caregiving and livelihood responsibilities more effectively without experiencing emotional burnout. Interventions in Senegal [60], Ghana [65], and Bangladesh [66] have also shown that MHPSS programs reduce anxiety and support women's psychological resilience.

The strategies employed by women in Wajir West demonstrate the central role of psychological resilience in their capacity to adapt to climate change. Peer support and social networks reduce isolation, faith-based coping provides spiritual grounding, and structured routines reduce mental strain. Self-care and community-based support enhance emotional well-being, enabling women to balance caregiving and livelihood activities. By maintaining emotional stability, women ensure household stability, support children's education, and maintain family well-being amid climate-induced shocks. Global evidence reinforces these findings. Studies in Senegal

[60, 67], Ghana [59, 68, 69] and Bangladesh [70–72] show that peer support networks, faith-based practices, and daily planning are key strategies for building women's psychological resilience to climate-induced stress. These approaches highlight the global relevance of women-led adaptation strategies for mental well-being. The policy implications are clear. Development actors should prioritize mental health as part of climate adaptation strategies. Expanding access to psychosocial support through savings groups and CBOs would strengthen women's emotional capacity to cope with climate shocks. Faith-based organizations should also be engaged, given the critical role of spiritual support in promoting mental well-being. Humanitarian organizations should integrate MHPSS services into climate adaptation programs to ensure women have access to self-care strategies, stress-relief techniques, and community support. By addressing the mental health needs of women, development actors can promote women's resilience, reduce mental health crises, and strengthen climate adaptation in vulnerable communities like Wajir West.

4.4 Theme 4: Disaster Risk Reduction and Emergency Response Strategies

Climate change in Wajir West sub-county has intensified the frequency and severity of natural disasters, particularly floods and droughts, disrupting livelihoods, displacing families, and destroying essential household resources. Women play a crucial role in disaster risk reduction (DRR) and emergency response due to their role as caregivers and household managers. They are often the first to recognize early warning signs and lead efforts in preparedness, evacuation, and recovery. Despite limited financial and technical resources, women employ a range of practical DRR strategies to safeguard their families and property.

A key strategy employed by women is household risk preparedness. Women store essential supplies such as food, water, firewood, and other basic necessities in elevated locations to ensure survival during floods. They also prepare emergency kits containing crucial items like clothes, documents, and food.

> "Before the floods, I prepare a bag with food, clothes, and important documents. When the water comes, I grab the bag and leave with my children." (Participant 2, individual interview)

Women also safeguard access to clean water by storing it in large containers ahead of floods.

> "I store water in large containers before the rainy season so that if the floodwaters mix with drinking water, we still have clean water to use." (Participant 8, interview)

These measures ensure that families have access to clean drinking water and essential supplies, reducing recovery time and enabling continuity of daily life even during displacement.

Emergency evacuation planning is another critical strategy. Women take charge of evacuation logistics, relocating children and vulnerable family members to safer areas as floods approach. Their ability to mobilize family members ahead of disasters reduces risks to life and property.

> "When I hear there is a flood warning, I tell my children to pack their clothes, and we all move to my sister's house, which is on higher ground." (Participant 6, focus group discussion)

Women also depend on social support networks, such as relatives and neighbors, to secure temporary shelter.

> "When the floods started, I moved with my children to a neighbor's house. We stayed there for two days until the water went down." (Participant 10, interview)

This role is critical in households where men have migrated in search of work, leaving women as the primary decision-makers.

Flood-proofing homes is another effective strategy. Women modify their homes by sealing wall cracks, raising furniture, and elevating critical household items to prevent flood damage.

> "I raised my bed and put wooden blocks under it so that if water enters the house, it does not soak the mattress." (Participant 12, interview)

Women also reinforce home structures using locally available materials like clay and mud.

> "We use mud to seal cracks in the walls before the rainy season. It keeps the water from entering the house." (Participant 7, focus group discussion)

These flood-proofing measures reduce financial strain from post-disaster repairs and enable families to continue with daily activities, such as caregiving and livelihood pursuits, without prolonged displacement.

Community-based preparedness systems further strengthen women's DRR strategies. Women engage in group planning, collective resource pooling, and financial preparedness. Through community meetings, they discuss evacuation plans, identify locations for storing food and water, and prepare financial resources for emergencies.

> "We meet as a group of women to discuss what to do before the floods come. We plan where to put our food and water and where to stay if we have to move." (Participant 11, interview)

Savings groups also play a critical role in financial preparedness, as women contribute emergency savings that can be accessed in times of crisis.

> "Our savings group agreed that each member should keep emergency money ready. When the floods came, I used my savings to buy food." (Participant 9, interview)

These community-based support systems enhance disaster response efficiency, reduce emergency response times, and ensure families have access to essential supplies during crises.

After floods subside, women lead the recovery and reconstruction of homes. Using funds from savings groups and support networks, they purchase materials to repair homes and restore normalcy.

> "After the floods, I used my savings from the Chama (savings group) to buy new roofing sheets for my house." (Participant 3, interview)

Women also engage in collective action with other community members, helping each other clean homes and remove mud from houses affected by floods.

> "The women in my group came together to help each other clean up after the floods. We removed mud from each other's homes and shared cleaning tools." (Participant 1, focus group discussion)

This collective action strengthens household recovery, shortens the time needed to restore normal living conditions, and enables women to resume caregiving and livelihood responsibilities.

Women's role in disaster risk reduction and emergency response is essential for climate adaptation in Wajir West. Their proactive actions in risk preparedness, evacuation, flood-proofing, and post-disaster recovery enhance household and community-wide resilience. Evidence from other regions supports these findings. In Ghana, women took on leadership roles in evacuation planning [73], while in Bangladesh, women used community support systems to strengthen flood responses [74]. In all cases, women's roles in evacuation, supply management, and community coordination were critical for reducing displacement, preserving livelihoods, and strengthening family resilience. Policymakers should prioritize financial inclusion for women, promote disaster preparedness training, and provide technical resources for flood-proofing homes. Empowering women with these resources will strengthen household resilience and promote community stability in the face of climate-induced shocks.

4.5 Theme 5: Health and Nutrition Strategies

Climate change in Wajir West sub-county has significantly affected access to healthcare and nutrition, especially during floods and droughts. These climate-induced shocks disrupt food production, reduce access to clean water, and increase disease outbreaks, posing severe challenges to household well-being. Women, as primary caregivers and household managers, play a pivotal role in addressing family health and nutrition. To mitigate these challenges, women have adopted a range of adaptation and resilience strategies to maintain household health during climate-induced disruptions.

One key strategy is the use of traditional herbal remedies. Women draw on indigenous knowledge to identify and use local medicinal plants to treat ailments like diarrhea, respiratory infections, and skin diseases. This approach reduces reliance on distant health facilities, which are often inaccessible due to floods or damaged roads.

> "When my child had a stomach problem, I used leaves from the 'Gob tree' to make medicine. It worked, and I didn't have to go to the health center." (Participant 2, individual interview)

Herbal remedies are also cost-effective, reducing the need for expensive medication.

> "We use plants like aloe vera and neem to treat skin problems and wounds. It saves us money because medicine from shops is too expensive." (Participant 5, focus group discussion)

These practices support women's ability to provide healthcare at home during climate-related health crises, reinforcing the importance of preserving indigenous knowledge. Similar strategies have been observed globally, with women in Ghana [75, 76] and Bangladesh [38, 77] using traditional herbs to address health issues during floods.

Another vital strategy is nutritional diversification. Climate-induced droughts and floods lead to food scarcity, forcing women to seek alternative food sources that are locally available, affordable, and drought-tolerant. Women increase their use of indigenous vegetables, wild fruits, and crops like cowpeas, millet, and sorghum.

> "When food runs out, we look for wild fruits like 'Matunda ya pori' (wild berries) to feed the children. It's not the best food, but it helps them stay full." (Participant 8, individual interview)

Women also switch to drought-tolerant crops to stabilize household food supplies.

> "During drought, we stop planting maize and start planting sorghum and millet because they can survive with little water." (Participant 3, interview)

By incorporating multiple food sources, women reduce their reliance on a single food supply, ensuring dietary balance and protecting children from malnutrition. Similar coping strategies have been noted in Ethiopia, where women diversified their diets using drought-tolerant crops like millet and sorghum [78].

Healthcare financing through savings groups is another critical strategy. Women pool resources into community savings groups, creating emergency funds that provide loans for healthcare services during floods and droughts. These funds are used to pay medical fees, purchase medicines, and finance transportation to healthcare centers.

> "When my son was sick, I borrowed money from our Chama to pay for his treatment. Without the group, I would have had to borrow from a shopkeeper at high interest." (Participant 4, individual interview)

Savings groups also provide timely support for medical emergencies, ensuring that treatment is not delayed.

> "We all contribute money, and when someone gets sick, she can borrow from the fund. Last year, I borrowed money to pay for my daughter's hospital fees." (Participant 7, focus group discussion)

This system offers women an alternative to high-interest loans from informal lenders, reducing financial strain on families. Similar systems have been observed in Bangladesh, where savings groups enabled women to finance healthcare during floods [72].

Psychosocial support mechanisms are also essential for mental well-being. Climate-induced crises place emotional strain on women as they manage caregiving, household responsibilities, and livelihood activities. Women rely on religious faith, peer support, and community counseling to maintain mental well-being.

> "When I feel overwhelmed, I pray and ask Allah for strength. It helps me stay calm." (Participant 1, interview)

Women also find emotional support through social interactions within savings groups.

> "We meet with other women at the savings group, and we talk about our problems. It helps because I know I'm not alone." (Participant 11, focus group discussion)

These strategies enable women to reduce stress and anxiety, allowing them to sustain their caregiving and livelihood responsibilities. Similar findings have been reported in Senegal, Ghana, and Bangladesh, where community support networks reduced emotional strain among women affected by climate-induced shocks [38, 59, 60].

The health and nutrition strategies adopted by women in Wajir West demonstrate their ability to adapt and sustain household well-being during climate-related shocks. Through herbal remedies, food diversification, savings group loans, and psychosocial support, women reduce dependence on external aid, increase household self-reliance, and protect family members from health-related impacts of climate change. Their strategies are consistent with those observed in Bangladesh, Ghana, and Ethiopia, highlighting the global relevance of women-led health resilience strategies. Policymakers should strengthen access to savings groups, support nutrition education, and facilitate access to climate-resilient healthcare systems. By supporting women's health and nutrition strategies, development actors can reduce health risks posed by climate change and enhance household resilience in vulnerable communities like Wajir West.

5 Conclusion

The study's findings underscore the central role of women in fostering resilience and adaptation strategies in response to climate change in Wajir West sub-county. The demographic profile of respondents revealed that most women were in their prime caregiving and working years, with limited access to formal employment and education. This backdrop shaped the adaptive strategies women employed to safeguard their households amid climate-induced disruptions. Key themes of resilience and adaptation emerged from the analysis. Women demonstrated economic adaptation through livelihood diversification, shifting from pastoralism to informal trading of firewood, vegetables, and milk. These efforts not only sustained household income but also highlighted women's evolving role as breadwinners. Community-based support networks, particularly women-led savings groups, emerged as vital pillars of resilience. These groups provided emergency loans for education, healthcare, and home repairs, ensuring that women could access financial resources during crises. The findings also revealed women's psychological resilience, marked by reliance on peer-based support, spiritual reflection, daily task planning, and access to mental health support. Disaster risk reduction strategies involved preparedness measures like stockpiling essential supplies, flood-proofing homes, and coordinating family evacuations. Lastly, health and nutrition strategies included the use of herbal remedies, dietary diversification through indigenous crops, healthcare financing through savings groups, and psychosocial support. Collectively, these strategies highlight the pivotal role of women as active agents of change. Their actions not only mitigate the immediate impacts of climate-induced shocks but also enhance household and community-wide resilience. Policymakers and development actors should strengthen women's access to financial, technical, and psychosocial support to foster sustainable and gender-sensitive climate adaptation.

5.1 Recommendations

Based on the study's findings, the following practical recommendations are proposed to strengthen women's resilience and adaptation strategies to climate change in Wajir West sub-county.

1. Livelihood diversification and economic empowerment: Development partners should facilitate vocational training programs for women in climate-resilient trades such as soap-making, tailoring, and poultry farming. Provision of start-up grants and in-kind support, such as equipment and raw materials, would enhance their capacity to engage in stable, alternative income-generating activities beyond informal trade.
2. Strengthening savings groups (chamas): Formalize and strengthen women-led savings groups by linking them to formal microfinance institutions. Capacity-building on financial literacy and record-keeping should be provided to ensure effective group management. Local government and NGOs can offer seed fund-

ing or matching grants to increase the capital base of these groups, enabling women to access larger loans for health, education, and livelihood support.

3. Mental health and psychosocial support (MHPSS): Partner with community-based organizations (CBOs) to offer regular mental health counseling sessions and support groups for women. Integrate stress management and self-care training into existing savings group activities, leveraging already established social support networks to provide emotional relief.
4. Disaster risk reduction (DRR) and emergency preparedness: Provide women with affordable flood-proofing kits, including water-resistant storage bags, raised platforms, and emergency lighting. Local authorities should strengthen early warning systems and train women in evacuation planning and safe relocation practices. Community storage hubs for essential supplies (food, water, and hygiene kits) should be established in flood-prone areas.
5. Health and nutrition resilience: Support women in cultivating drought-tolerant crops (like millet and sorghum) through access to seeds, training, and small irrigation kits. Local health departments and NGOs should promote indigenous herbal remedies by training women on safe and effective use. Community health insurance schemes linked to savings groups should be established to provide affordable healthcare access during emergencies.

References

1. Mondal, D. (2024). Socio-economic impacts of climate change. *International Journal on Agricultural Sciences, 15*(2), 95–96. https://doi.org/10.53390/IJAS.2024.15203
2. Rocha, J., Oliveira, S., Viana, C. M., & Ribeiro, A. I. (2022). Chapter 8—Climate change and its impacts on health, environment and economy. In J. C. Prata, A. I. Ribeiro, & T. Rocha-Santos (Eds.), *One Health* (pp. 253–279). Academic Press. https://doi.org/10.1016/B978-0-12-822794-7.00009-5
3. Choudhary, M. (2024). Gender frontlines: Exploring the intersectionality of gender and climate change. *BSSS Journal of Social Work, XVI*(1), 1–9. https://doi.org/10.51767/jsw1601
4. John, S. (2025). The unequal burden: Climate change and vulnerable populations. In *Gender, environment, and human rights: An intersectional exploration* (pp. 359–380). IGI Global.
5. Abid, Z., Abid, M., Zafar, Q., & Mehmood, S. (2018). Detrimental effects of climate change on women. *Earth Systems and Environment, 2*(3), 537–551. https://doi.org/10.1007/s41748-018-0063-9
6. Elizabeth, A. (2017). Toward a feminist care ethic for climate change. *Journal of Feminist Studies in Religion, 33*(2), 152–158. https://doi.org/10.2979/jfemistudreli.33.2.15
7. Spitzer, P. A., & Spitzer, P. A. (2023). Why a female-centered approach? In *Empowering female climate change activists in the global south: The path toward environmental social justice*. Emerald Publishing Limited. https://doi.org/10.1108/978-1-80382-919-720231002
8. Lewis, N. D. (2016). Sustainable development through a gendered lens: Climate change adaptation and disaster risk reduction. *Reviews on Environmental Health, 31*(1), 97–102. https://doi.org/10.1515/reveh-2015-0077
9. Ibe, G., & Amikuzuno, J. (2019). Climate change in Sub-Saharan Africa: A menace to agricultural productivity and ecological protection. *Journal of Applied Sciences and Environmental Management, 23*(2), 329–335. https://doi.org/10.4314/jasem.v23i2.20

10. Kula, N., Haines, A., & Fryatt, R. (2013). Reducing vulnerability to climate change in sub-Saharan Africa: the need for better evidence. *PLoS Medicine, 10*(1), e1001374. https://doi.org/10.1371/journal.pmed.1001374
11. Serdeczny, O., Adams, S., Baarsch, F., Coumou, D., Robinson, A., Hare, W., Schaeffer, M., Perrette, M., & Reinhardt, J. (2017). Climate change impacts in Sub-Saharan Africa: From physical changes to their social repercussions. *Regional Environmental Change, 17*(6), 1585–1600.
12. Tofu, D. A., Fana, C., Dilbato, T., Dirbaba, N. B., & Tesso, G. (2023). Pastoralists' and agro-pastoralists' livelihood resilience to climate change-induced risks in the Borana zone, south Ethiopia: Using resilience index measurement approach. *Pastoralism, 13*(1), 4. https://doi.org/10.1186/s13570-022-00263-3
13. Kalele, D. N., Ogara, W. O., Oludhe, C., & Onono, J. O. (2021). Climate change impacts and relevance of smallholder farmers' response in arid and semi-arid lands in Kenya. *Scientific African, 12*, e00814.
14. Korir, J. C. (2019). Climate change stress on pastoral livelihoods in Narok County, Kenya. *European Journal of Social Sciences Studies*.
15. Wanjara, A. O., & Ogembo, P. O. (2023). Impact of climate change on health and livelihoods of pastoral communities in Kenya: A case of north eastern region. *Indonesian Journal of Social and Environmental Issues (IJSEI), 4*(3), 299–315. https://doi.org/10.47540/ijsei.v4i3.1078
16. Allen, N. (2022). Gender disparity and climate change—Addressing the disproportionate effects of climate change on women. *Global Energy Law and Sustainability, 3*(2), 206–226. https://doi.org/10.3366/gels.2022.0080
17. Pharswan, S., Pipaliya, U., & Singh, R. (2025). A systematic review on biopsychosocial impact of climate change on women's health and well-being. In J. Kurebwa & D. Makwerere (Eds.), *Managing the health risks of climate change* (pp. 1–32). IGI Global Scientific Publishing. https://doi.org/10.4018/979-8-3693-4183-4.ch001
18. Akuja, T. E., & Kandagor, J. J. (2024). Climate smart agriculture in Kenya's ASALS: Gaps and barriers in policy development and implementation. *African Journal of Climate Change and Resource Sustainability, 3*(1). https://doi.org/10.37284/ajccrs.3.1.1690
19. Nunow, A. A. (2024). The nexus between climate change and livelihoods in arid and semi-arid (ASAL) areas of Kenya: Evidence from the literature. *International Journal of Social Science and Human Research, 07*(06), 4500–4509. https://doi.org/10.47191/ijsshr/v7-i06-113
20. Ouma, J. O., Olang, L. O., Ouma, G. O., Oludhe, C., Ogallo, L., & Artan, G. (2018). Magnitudes of climate variability and changes over the arid and semi-arid lands of Kenya between 1961 and 2013 period. *American Journal of Climate Change, 7*(01), 27.
21. Ochola, E., Otieno, M., & Okebe, M. (2022). The impact of floods on Kenyan Women: A critical review of media coverage, institutional response and opportunities for gender responsive mitigation. *Nairobi Journal of Humanities and Social Sciences, 6*(1), 17–32. https://doi.org/10.58256/njhs.v6i1.732
22. Opondo, D. O. (2013). Erosive coping after the 2011 floods in Kenya. *International Journal of Global Warming, 5*(4), 452–466.
23. Oxford, A. (2024). Floods add to pressure on Kenya's government. *Expert Briefings*. https://doi.org/10.1108/OXAN-ES286907
24. Weesie, R. V., Rohse, M., Van Loon, A. F., Koehler, J., Barendrecht, M. H., & Mwangi, M. (2025). Floods after drought: Storytelling with agro-pastoralists in a Kenyan dryland. *Frontiers in Water, 7*, 1524554. https://doi.org/10.3389/frwa.2025.1524554
25. Karmakar, S. (2021). Gender and climate change: The condition of women in the Indian Sundarbans. *Scholars Journal of Arts, Humanities and Social Sciences, 1*(1), 1–5. https://doi.org/10.36347/sjahss.2021.v09i01.001
26. Sahiwal, T. K., & Gupta, J. K. (2025). Women at the crossroads of climate change: Advancing gender equality through climate action. In S. Chakraborty & S. S. Satapathy (Eds.), *Gender, environment, and human rights: An intersectional exploration* (pp. 177–210). IGI Global Scientific Publishing. https://doi.org/10.4018/979-8-3693-6069-9.ch011

27. Adeola, O., Evans, O., & Ngare, I. (2024). Gender equality, climate action, and technological innovation for sustainable development in Africa: Conclusion and recommendations. In O. Adeola, O. Evans, & I. Ngare (Eds.), *Gender equality, climate action, and technological innovation for sustainable development in Africa* (pp. 209–241). Springer. https://doi.org/10.1007/978-3-031-40124-4_10
28. Rezk, H., & Sayed, E. T. (2024). Bibliometric analysis highlighting the role of women in addressing climate change challenges and achieving sustainable development goals for greener future. *International Journal of Electrical & Computer Engineering (2088-8708), 14*(3).
29. Sidun, N. M., & Gibbons, J. L. (2024). Women, girls, and climate change: Human rights, vulnerabilities, and opportunities. *International Journal of Psychology, 59*(2), 257–266.
30. Alam, K., & Rahman, M. H. (2018). The role of women in disaster resilience. In *Handbook of disaster risk reduction & management* (pp. 697–719). World Scientific.
31. Schilling, J., & Werland, L. (2023). Facing old and new risks in arid environments: The case of pastoral communities in Northern Kenya. *PLoS Climate, 2*(7), e0000251.
32. Tadele, F., & Manyena, B. S. (2009). Building disaster resilience through capacity building in Ethiopia. *Disaster Prevention and Management: An International Journal, 18*(3), 317–326.
33. Atela, J., Gannon, K. E., & Crick, F. (2018). *Climate change adaptation among female-led micro, small, and medium enterprises in semiarid areas: A case study from Kenya* (pp. 1001–1018). Springer. https://doi.org/10.1007/978-3-319-71025-9_97-1
34. Ngigi, M. W., Mueller, U., & Birner, R. (2016). Gender differences in climate change perceptions and adaptation strategies: An intra-household analysis from rural Kenya. *Social Science Research Network*, 1–34. https://doi.org/10.22004/AG.ECON.232900
35. Chambers, R., & Conway, G. (1992). *Sustainable rural livelihoods: practical concepts for the 21st century.*
36. Scoones, I. (1998). *Sustainable rural livelihoods: a framework for analysis.*
37. Dadzie, S. K. N., Inkoom, E. W., Akaba, S., Annor-Frempong, F., & Afful, J. (2021). Sustainability responses to climate-smart adaptation in Africa: Implication for food security among farm households in the Central Region of Ghana. *African Journal of Economic and Management Studies, 12*(2), 208–227.
38. Huq, H., Rahman, M. M., & Hossen, M. A. (2023). Women-led climate change adaptation: A study in northwestern Bangladesh. *Sustainability*. https://doi.org/10.3390/su152015041
39. Tschakert, P., & Machado, M. (2012). Gender justice and rights in climate change adaptation: Opportunities and pitfalls. *Ethics and Social Welfare, 6*(3), 275–289.
40. Bryan, E., Ringler, C., & Meinzen-Dick, R. (2023). Gender, resilience, and food systems. In C. Béné & S. Devereux (Eds.), *Resilience and food security in a food systems context* (pp. 239–280). Springer. https://doi.org/10.1007/978-3-031-23535-1_8
41. DFID. (1999). *Sustainable livelihoods guidance sheets.* Department for International Development.
42. Mohan, R., Menon, P., Anjana, J., & Aswathy, J. (2024). *Qualitative assessment of resilience: Lessons from rural women in South India.* International Conference on Gender Research.
43. Parks-Yancy, R., DiTomaso, N., & Post, C. (2007). The mitigating effects of social and financial capital resources on hardships. *Journal of Family and Economic Issues, 28*(3), 429–448. https://doi.org/10.1007/s10834-007-9065-8
44. Nagasha, J. I., Ocaido, M., & Kaase-Bwanga, E. (2019). Theoretical and conceptual framework for gender analysis of attitudes and adaptation mechanisms to climate change for sustainable livelihoods in Uganda. *Journal of African Studies and Development, 11*(4), 51–58. https://doi.org/10.5897/JASD2019.0532
45. Kast, F., Meier, S., & Pomeranz, D. (2012). *Under-savers anonymous: Evidence on self-help groups and peer pressure as a savings commitment device.*
46. Mersland, R., & Eggen, Ø. (2007). You cannot save alone-financial and social mobilization in savings and credit groups. *Available at SSRN 1032247.*
47. Pelling, M. (2010). *Adaptation to climate change: From resilience to transformation.* Routledge.

48. Ahmad, N., El-Fatal, L., Pehu, E., Poutiainen, P., & Vyzaki, M. (2014). *Gender, agriculture, and climate change*. 1–4. https://documents.worldbank.org/curated/en/500281468154786516/Gender-agriculture-and-climate-change
49. Awuviri, S. (2017). *A study on how the creation of new value chains for Shea butter production influences the livelihood of rural women in a climate change situation in Northern Ghana*. https://thesis.eur.nl/pub/41764/SAwuviri_MA_2016_17_AFES.pdf
50. Lecoutere, E., Mishra, A., Singaraju, N., Koo, J., Azzarri, C., Chanana, N., Nico, G., & Puskur, R. (2023). Where women in agri-food systems are at highest climate risk: A methodology for mapping climate–agriculture–gender inequality hotspots. *Frontiers in Sustainable Food Systems*. https://doi.org/10.3389/fsufs.2023.1197809
51. Addo, P., Djekic-Ivankovic, M., Abdu, A., Boadi, P., Eyo, O., Baguignan, M., Atakoun, A., Adamagnon, E., Oso, O., & Owusu, J. (2024). Gender role in food rights and sovereignty in West Africa: A rapid review. *African Journal of Food, Agriculture, Nutrition and Development, 24*(10), 24725–24773.
52. Ndumbe, L. N. (2013). 3 Unshackling women traders: Cross-border trade of Eru from Cameroon to Nigeria. In *Women and trade in Africa: Realizing the potential* (p. 43).
53. Chowdhury, M., Masud, M., Kachna, P. K., & Rangpur, B. (2020). Impact of flood induced migration on livelihood and gender relation: A study on Chilmari, Kurigram. *International Journal of Engineering Applied Sciences and Technology, 5*(5), 1–7.
54. Haile, E. T. (2008). *Gender role and pastoralist women's involvement in income generating activities* [Wageningen UR].
55. Scheurlen, E. (2015). *Time allocation to energy resource collection in rural Ethiopia: Gender-disaggregated household responses to changes in firewood availability*.
56. Zoynul, A. M., & Fahmida, M. (2013). Women empowerment through micro credit: A case study of Dinajpur, Bangladesh. *Studies in Business & Economics, 8*(2).
57. Berg, F. (2017). *Building women's disaster resilience: An investigation of social capital generation through international disaster assistance following cyclone pam*.
58. Lukwa, A. T., Odunitan-Wayas, F., Lambert, E. V., Alaba, O. A., Collaborators, S., & f. H. I. (2022). Can informal savings groups promote food security and social, economic and health transformations, especially among women in urban Sub-Saharan Africa: A narrative systematic review. *Sustainability, 14*(6), 3153.
59. Dapilah, F., Nielsen, J. Ø., & Friis, C. (2020). The role of social networks in building adaptive capacity and resilience to climate change: A case study from northern Ghana. *Climate and Development, 12*(1), 42–56. https://doi.org/10.1080/17565529.2019.1596063
60. Diouf, A. C., Cissokho, D., & Somadjago, M. (2020). *Analyse des stratégies de résilience développées par les femmes du DELTA DU SALOUM (SENEGAL) face aux changements climatiques. 32*. https://doi.org/10.34874/IMIST.PRSM/EGSM/19375
61. Rahman, F. (2021). 'I find comfort here': Rohingya women and taleems in Bangladesh's refugee camps. *Journal of Refugee Studies, 34*(1), 874–889. https://doi.org/10.1093/jrs/fez054
62. Yessoufou, A. N.-D., Kumar, S., Houessionon, P., Worou, O. N., Wane, A., & Whitbread, A. (2024). Vulnerability and resilience in the face of climate changes in Senegal's drylands: measurement at the household level and determinant assessment. *Frontiers in Climate*. https://doi.org/10.3389/fclim.2024.1330025
63. Acheampong, P. P., Yeboah, S., Adabah, R., Asibuo, J., Nchanji, E. B., Opoku, M., Toywa, J., & Lutomia, C. K. (2023). Gendered perceptions and adaptations to climate change in Ghana: What factors influence the choice of an adaptation strategy? *Frontiers in Sustainable Food Systems, 7*. https://doi.org/10.3389/fsufs.2023.1091812
64. Kabeer, N., & Huq, L. (2010). The power of relationships: Love and solidarity in a landless women's organisation in rural Bangladesh. *IDS Bulletin, 41*(2), 79–87.
65. Assan, E., Suvedi, M., Olabisi, L. S., & Allen, A. (2018). Coping with and adapting to climate change: A gender perspective from smallholder farming in Ghana. *Environments, 5*(8), 86. https://doi.org/10.3390/ENVIRONMENTS5080086

66. Goudet, J.-M., Owais, H. M. A., Ahmed, H. U., & Ridde, V. (2023). Intersecting vulnerabilities: Unraveling the complex relationship between women, climate change, and mental health in Bangladeshi communities. *medRxiv*. https://doi.org/10.1101/2023.06.06.23290721
67. Sy, I. (2024). *Community resilience: integrating local approaches into Senegal's climate change adaptation agenda*. https://doi.org/10.59184/pb023.08
68. Antwi-Agyei, P., & Nyantakyi-Frimpong, H. (2021). Evidence of climate change coping and adaptation practices by smallholder farmers in Northern Ghana. *Sustainability, 13*(3), 1308. https://doi.org/10.3390/SU13031308
69. Gyan, C., & Kwakye, J. (2024). Building resiliency in community development: The experiences of women in rural communities in Ghana. *Social Inclusion, 13*. https://doi.org/10.17645/si.8705
70. Goudet, J.-M., Arif, F. B., Owais, H., Ahmed, H. U., & Ridde, V. (2024). Climate change and women's mental health in two vulnerable communities of Bangladesh: An ethnographic study. *PLOS Global Public Health, 4*(6), e0002080. https://doi.org/10.1371/journal.pgph.0002080
71. Lilier, K., Selim, S. A., Raihan, S., Islam, R., Das, J., Danquah, I. B., Sauerborn, R., & Bärnighausen, K. (2022). Coping strategies and barriers to coping in climate-vulnerable Bangladesh: A qualitative study. *European Journal of Public Health, 32*(Supplement_3). https://doi.org/10.1093/eurpub/ckac131.160
72. Moni, S. A. (2024). Empowering women in climate-resilient adaptation: A women, peace, and security (WPS) approach to addressing climate challenges in Bangladesh. *International Journal of Science and Research Archive, 13*(2), 1764–1776. https://doi.org/10.30574/ijsra.2024.13.2.2339
73. Ersing, R. L., & Caruson, K. (2017). Strengthening hazard resilience: Building capacity by empowering marginalized women. *International Journal of Mass Emergencies & Disasters, 35*(2), 49–69. https://doi.org/10.1177/028072701703500204
74. Datta, R., Kairy, B., & Hurlbert, M. (2024). Addressing wetland flood disasters through community-led strategies in Bangladesh. *Asia-Pacific Journal of Rural Development, 34*(1), 98–112. https://doi.org/10.1177/10185291241296587
75. Sedegah, D. D., Akagbor, M. A., Kwarteng-Donkor, P. B., Adu-Okoree, B. I., & Premkumar, P. J. J. (2023). Exploring the praxes of indigenous and experiential knowledge of climate change on peasant farmers in Ghana. *Journal of Sustainability Science and Management, 18*(10), 70–80. https://doi.org/10.46754/jssm.2023.10.005
76. Yarney, L., Sakyi, E., & Achamwie, P. (2021). *Climate change and rural female farmers in Ghana: A study of the Wenchi Municipality*. https://doi.org/10.20944/PREPRINTS202106.0088.V1
77. Garai, J., & Ku Hok-Bun, B. (2024). Cultural adaptation in the era of climate change: An ethnographic study on the resilience of indigenous people at Chittagong Hill Tracts area in Bangladesh. *Natural Hazards Research*. https://doi.org/10.1016/j.nhres.2024.05.002
78. Desta, D. T., Teferra, T. F., & Gebremedhin, S. (2024). Characterisation of complementary feeding practice and locally available climate-resilient crops for complementary food among agro-pastoralists of Ethiopia: A qualitative study. Journal of Nutritional Science, 13, e39. https://doi.org/10.1017/jns.2024.53

Index

N. Ronoh, C. Rono (eds.), *Rethinking Our Ecosystems*, Springer Proceedings in Geography and Sustainability, https://doi.org/10.1007/978-3-031-95781-9

MIX
Papier aus verantwortungsvollen Quellen
Paper from responsible sources
FSC® C105338

If you have any concerns about our products,
you can contact us on
ProductSafety@springernature.com

In case Publisher is established outside the EU,
the EU authorized representative is:
Springer Nature Customer Service Center GmbH
Europaplatz 3, 69115 Heidelberg, Germany

Printed by Libri Plureos GmbH
in Hamburg, Germany